Alisher Rahimzoda

Asymptotic formulas in the Esterman problem

Alisher Rahimzoda

Asymptotic formulas in the Esterman problem

Fourth degree with almost equal terms

ScienciaScripts

Imprint

Any brand names and product names mentioned in this book are subject to trademark, brand or patent protection and are trademarks or registered trademarks of their respective holders. The use of brand names, product names, common names, trade names, product descriptions etc. even without a particular marking in this work is in no way to be construed to mean that such names may be regarded as unrestricted in respect of trademark and brand protection legislation and could thus be used by anyone.

Cover image: www.ingimage.com

This book is a translation from the original published under ISBN 978-620-7-81055-0.

Publisher:
Sciencia Scripts
is a trademark of
Dodo Books Indian Ocean Ltd. and OmniScriptum S.R.L publishing group

120 High Road, East Finchley, London, N2 9ED, United Kingdom
Str. Armeneasca 28/1, office 1, Chisinau MD-2012, Republic of Moldova, Europe
Printed at: see last page
ISBN: 978-620-7-92247-5

Copyright © Alisher Rahimzoda
Copyright © 2024 Dodo Books Indian Ocean Ltd. and OmniScriptum S.R.L publishing group

Institute of Mathematics named after A.Dzhurayev. A. Juraev

Institute of Mathematics

National Academy of Sciences of Tajikistan

RAHIMZODA ALISHER ORZU

ASYMPTOTIC FORMULA IN THE ESTERMAN PROBLEM OF DEGREE FOUR WITH ALMOST EQUAL SUMMANDS

Dushanbe - 2024

TABLE OF CONTENTS

2

Designations

$$e(\alpha) = e^{2\pi i \alpha} = \cos 2\pi\alpha + i\sin 2\pi\alpha.$$

In references, theorems, lemmas, and formulas are numbered with three indices: chapter number, paragraph number, and statement number.

$c, c_1, c_2, \cdots,$-positive constants, not always the same ones.

ε-positive any small constants.

$\varphi(q)$ - Euler function.

$\mu(n)$ - Möbius function.

$\Lambda(n)$ - Mangoldt function.

$\tau(n)$ - number of divisors of the number n.

$\tau_r(n)$ - number of solutions to the equation $x_1 x_2 \ldots x_r = n$ in natural numbers $x_1, x_2, \ldots, x_r$.

Recording $A \asymp B$ means that $c_1 A \leq B \leq c_2 A$.

In the affirmative. A entry $B = O(A)$ or $B \ll A$ means that there exists $c > 0$ such that $|B| \leq cA$.

(a, b) - greatest common divisor of the numbers a и b.

$[x]$ - integer part x.

$\{x\}$ - fractional part x.

$\| x \| = \min(\{x\}, 1 - \{x\})$ - distance to the nearest integer.

$\mathcal{L} = \ln xq.$

<u>Introduction</u>

The present paper is a study in analytic number theory of short trigonometric sums and its applications to classical additive problems with more rigid conditions, namely, when the summands are almost equal. Historically, the following studies were the first examples of the study of such problems:

- Short trigonometric sums, which arise when solving additive problems with almost equal summands, were first studied by I. M. Vinogradov [1]. M. Vinogradov [1]. He first studied for a short linear trigonometric sum with simple numbers, i.e. for sums of the form:

$$S_k(\alpha; x, y) = \sum_{x-y < n \leq x} \Lambda(n) e(\alpha n^k), \qquad \alpha = \frac{a}{q} + \lambda, \qquad |\lambda|$$

$$\leq \frac{1}{q\tau}, \qquad 1 \leq q \leq \tau,$$

at $k = 1$ using his method of evaluations of sums with prime numbers, proved a nontrivial evaluation at

$$\exp(c(\ln\ln x)^2) \ll q \ll x^{1/3}, \qquad y > x^{2/3+\varepsilon}.$$

Then S. B. Haselgrove [2] obtained a nontrivial estimate of the sum of $S_1(\alpha; x, y)$, $y \geq x^{\theta}$, q - arbitrary, and proved an asymptotic formula for the ternary Goldbach problem with almost equal summands, i.e. for the number of solutions of the Diophantine equation

$$N = p_1 + p_2 + p_3, \tag{1}$$

in terms of

$$\left| p_i - \frac{N}{3} \right| \leq H, \qquad H = N^{\theta+\varepsilon}, \qquad \theta = \frac{63}{64} + \varepsilon. \tag{2}$$

This was the first solved additive problem with almost equal summands. Then V. Statuliavychus [3], J. Chaohua [4, 5, 6, 7], Pan Chen-dong and Pan Chen-biao [8], and T. Zhang [9] replaced the exponent θ respectively by

$$\frac{279}{308} + \varepsilon, \qquad \frac{91}{96} + \varepsilon, \qquad \frac{13}{17} + \varepsilon, \qquad \frac{2}{3} + \varepsilon, \qquad \frac{5}{8} + \varepsilon.$$

The best result in this problem belongs to J. Chaohua [10]. He proved that the Diophantine equation (1) with conditions (2) is solvable with exponent

$$\theta = \frac{7}{12} + \varepsilon.$$

- T. Zhang and J. Liu [11, 12, 13, 14], having obtained a nontrivial estimate of the sum of $S_2(\alpha, x, y)$ They proved Hua Lo Geng's theorem on the representability of a sufficiently large natural number N, $N \equiv 5 (mod\, 24)$ as a sum of five squares of prime numbers in the case when these summands are almost equal. They showed that a sufficiently large natural number N, $N \equiv 5 (mod\, 24)$, can be represented as

$$N = p_1^2 + \cdots + p_5^2, \qquad \left| p_j - \sqrt{\frac{N}{5}} \right| \le H, \qquad H \ge N^{\frac{11}{23}+\varepsilon}.$$

T. Jean and J. Liu [13], taking advantage of the obtained sum estimation $S_2(\alpha, x, y)$ also showed that a sufficiently large natural number N can be represented in the form $N = p_1 + p_2 + p_3^2$ with the conditions

$$\left| p_i - \frac{N}{3} \right| \le H, \qquad i = 1,2, \qquad \left| p_3^2 - \frac{N}{3} \right| \le H, \qquad H \ge N^{\frac{11}{23}+\varepsilon}.$$

- In 1938, Hua [15], considering the Waring-Goldbach problem for cubes, proved that all sufficiently large odd natural numbers are the sum of nine cubes of prime numbers. A. V. Kumchev [16] obtained a nontrivial estimate of the sum of $S_k(\alpha; x, y)$ in small arcs $m(P)$ at $y \geq x^{\theta+\varepsilon}$, $\theta = 1 - \frac{1}{2k+3}$ и $\tau = x^{1+2\theta}P^{-1}$. Ya Yao. [17], using Kumchev's estimate, proved that any sufficiently large odd natural number N can be represented in the form

$$p_1^3 + p_2^3 + \cdots + p_9^3 = N, \qquad \left| p_i - \sqrt[3]{\frac{N}{9}} \right| \leq N^{\frac{1}{3} - \frac{1}{51} + \varepsilon}.$$

- Short trigonometric sums of G. Weyl of the form

$$T(\alpha, x, y) = \sum_{x-y < m \leq x} e(\alpha m^n),$$

at $n = 2,3,4$ in long arcs, were studied in [18, 19, 20, 21, 22]. These results were applied in the derivation of asymptotic formulas in the following additive problems with almost equal summands:

in Esterman's problem [18, 23] on the representation of a natural number $N > N_0$ in the form $p_1 + p_2 + m^2 = N$, p_1 и p_2 - prime numbers, $m > 0$ - integer, with the conditions

$$\left| p_i - \frac{N}{3} \right| \leq H; \qquad i = 1,2, \qquad \left| m^2 - \frac{N}{3} \right| \leq$$

$H,$ $\qquad H \geq N^{\frac{3}{4}}\ln^2 N;$

in the cubic Esterman problem [20, 24] on the representation of a natural number $N > N_0$ in the form $p_1 + p_2 + m^3 = N$, p_1 и p_2 - prime numbers, $m > 0$ - integer, with the conditions

$$\left| p_i - \frac{N}{3} \right| \leq H; \qquad i = 1,2, \qquad \left| m^3 - \frac{N}{3} \right| \leq$$

$H,$ $\qquad H \geq N^{\frac{5}{6}}\ln^3 N;$

in the Varing problem for cubes [25] on the representation of a natural number $N > N_0$ as nine cubes of natural numbers x_i, $i = \overline{1,9}$ with the conditions

$$\left| x_i - \left(\frac{N}{9} \right)^{\frac{1}{3}} \right| \leq H, \qquad H \geq N^{\frac{1}{3} - \frac{1}{30} + \varepsilon};$$

in the Varing problem for fourth powers [26] on the representation of a natural number $N > N_0$ as a sum of seventeen fourth powers of natural numbers x_i, $i = \overline{1,17}$ with the conditions

$$\left| x_i - \left(\frac{N}{17} \right)^{\frac{1}{4}} \right| \leq H, \qquad H \geq N^{\frac{1}{4} - \frac{1}{108} + \varepsilon};$$

in the Varing problem for fifth powers [27] on the representation of a natural number $N > N_0$ as a sum of 33 fifth powers of natural numbers x_i, $i = \overline{1,33}$ with the conditions

$$\left| x_i - \left(\frac{N}{33} \right)^{\frac{1}{5}} \right| \leq H, \qquad H \geq N^{\frac{1}{5} - \frac{1}{340} + \varepsilon}.$$

- S.Y. Fatkina [28] proved an asymptotic formula for the number of solutions of the Diophantine equation $N = p_1 + p_2 + [\sqrt{2}p_3]$ in prime numbers p_1, p_2, p_3, with the conditions

$$\left| p_i - \frac{N}{3} \right| \leq H, \qquad i = 1,2, \qquad \left| [\sqrt{2}p_3] - \frac{N}{3} \right| \leq H, \qquad H \geq$$

$N^{\frac{5}{8}}\ln^c N.$

- P.Z. Rakhmonov [29, 30] at $y \gg \sqrt{x}$ obtained a uniform estimate on the parameter c for short trigonometric sums with non-integer degree of a natural number of the form

$$S_c(\alpha; x, y) = \sum_{x-y<n\leq x} e(\alpha[n^c]),$$

and proved an asymptotic formula in the generalization of the ternary Esterman problem for non-integer degrees with almost equal summands on the representation of a sufficiently large natural number as $p_1 + p_2 + [n^c] = N$ in prime numbers p_1, p_2 and natural n, with conditions at

$$\left|p_i - \frac{N}{3}\right| \leq H, \qquad i = 1,2, \qquad \left|[n^c] - \frac{N}{3}\right| \leq H, \qquad H \geq$$

$N^{1-\frac{1}{2c}}\ln^2 N.$

The relevance and appropriateness of this monograph is determined by the fact that it

     - the behavior of short trigonometric sums of G. Weyl of the form

$$T(\alpha, x, y) = \sum_{x-y<m\leq x} e(\alpha m^n),$$

in large arcs;

     - the obtained results allowed us to find an asymptotic formula for the number of representations of a sufficiently large natural number as a sum of three almost equal summands, two of which are prime numbers and the third is the fourth degree of the natural number. It should be noted that the sequence under consideration is rarer than those mentioned above.

Purpose of the work.

The aim of this paper is to study the behavior of short Weyl trigonometric sums in large arcs, to evaluate such fourth-order sums in small arcs and their applications to the derivation of an asymptotic formula in the Esterman problem of degree four with almost equal summands.

Research **Methods**

The degree of validity of the scientific results obtained in the thesis is confirmed by rigorous mathematical proofs obtained by applying modern methods of analytic number theory, namely:

- method of estimation of special trigonometric sums and van der Korput integrals with application of Poisson summation formula, estimation of trigonometric integrals by the modulus of derivatives, estimation of full rational sums Hua Lo-Ken;

- G. Weil's method of estimations of trigonometric sums;

- The circular method of Hardy, Littlewood and Ramanujan in the form of trigonometric sums I.M. Vinogradov.

Scientific Novelty.

The main results of the thesis are novel, they are substantiated by detailed evidence and are as follows:

- the behavior of short trigonometric sums of G. Weyl in large arcs is studied;

- a nontrivial estimate of short Weyl trigonometric sums of fourth order in small arcs is found;

- we prove an asymptotic formula for the number of representations of a sufficiently large natural number as a sum of three almost equal summands, two of which are prime numbers and the third is the fourth degree of the natural number.

Theoretical and practical value

The monograph has a theoretical character. Its results and the methodology of their derivation can be used by specialists in the field of analytic number theory.

Structure and scope ё Scope of work. The work consists of an introduction, two chapters divided into paragraphs. The total volume of the work is 57 pages. The list of cited literature includes 53 names.

Summary of work

The monograph consists of an introduction and two chapters. The introduction of the work contains an overview of the results related to the topic of the work, and also formulates the main results obtained in it.

The first paragraph of the first chapter is of an auxiliary nature, where known lemmas are given, which are applied in the following paragraphs.

P. Vaughn [31], studying G. Weyl sums of the form

$$T(\alpha, x) = \sum_{m \leq x} e(\alpha m^n), \qquad \alpha = \frac{a}{q} + \lambda, \ q \leq$$

$$\tau, \qquad (a, q) = 1, \ |\lambda| \leq \frac{1}{q\tau},$$

in large arcs using the van der Korput method:

$$T(\alpha, x) = \frac{S(a,q)}{q} \int_0^x e(\lambda t^n) dt + O\left(q^{\frac{1}{2}+\varepsilon}(1 + x^n|\lambda|)^{\frac{1}{2}}\right).$$

Provided that α is very well approximated by a rational number with denominator q that is, if the condition

$$|\lambda| \leq \frac{1}{2nqx^{n-1}},$$

he also proved

$$T(\alpha, x) = \frac{x}{q} \cdot \frac{S(a,q)}{q} \int_0^1 e(\lambda t^n)dt + O\left(q^{\frac{1}{2}+\varepsilon}\right).$$

When deriving asymptotic formulas in additive problems with almost equal summands, which include the Varing problem and the Esterman problem, the main point, along with the circular Hardy-Littlewood method in the variant of trigonometric sums of I. M. Vinogradov, is also the behavior of short trigonometric sums of G. Weyl of the form

$$T(\alpha; x, y) = \sum_{x-y<m\leq x} e(\alpha m^n), \quad \alpha = \frac{a}{q} + \lambda, \qquad (a, q) =$$

1, $\quad q \leq \tau, \ |\lambda| \leq \frac{1}{q\tau},$

in large arcs and their evaluation in small arcs.

Behavior $T(\alpha; x, y)$ in large arcs is studied in the second paragraph of the first chapter and the main result is **Theorem 1.2**, which simplifies the proof and clarifies the main theorem of [32, 33].

Theorem 1.2. *Let* $\tau \geq 2n(n-1)x^{n-2}y$ *и* $\lambda \geq 0$, *then at* $\{n\lambda x^{n-1}\} \leq \frac{1}{2q}$, *the formula*

$$T(\alpha, x, y) = \frac{S(a,q)}{q} T(\lambda; x, y) + O(q^{\frac{1}{2}+\varepsilon}),$$

and at $\{n\lambda x^{n-1}\} > \frac{1}{2q}$ there is an estimation

$$|T(\alpha, x, y)| \ll q^{1-\frac{1}{n}}\ln q + \min_{2\leq k\leq n} \left(yq^{-\frac{1}{n}}, \lambda^{-\frac{1}{k}}x^{1-\frac{n}{k}}q^{-\frac{1}{n}}\right).$$

Corollary 1.2. *Let* $\tau \geq 2n(n-1)x^{n-2}y$, $|\lambda| \leq \frac{1}{2nqx^{n-1}}$ *then the*

relation

$$T(\alpha, x, y) = \frac{y}{q} S(a, q)\gamma(\lambda; x, y) + O(q^{\frac{1}{2}+\varepsilon}).$$

11

Corollary 1.3. *Let* $\tau \geq 2n(n-1)x^{n-2}y$, $\dfrac{1}{2nqx^{n-1}} < |\lambda| \leq \dfrac{1}{q\tau}$,

then the estimation

$$T(\alpha, x, y) \ll q^{1-\frac{1}{n}}\ln q + \min_{2 \leq k \leq n}\left(yq^{-\frac{1}{n}}, x^{1-\frac{1}{k}}q^{\frac{1}{k}-\frac{1}{n}}\right).$$

Corollaries 1.2 and 1.3 are generalizations of the above results of R. Vaughn for short trigonometric sums of G. Weyl $T(\alpha, x, y)$.

The proof of Theorem 2 is carried out by the method of evaluation of special trigonometric sums by van der Korput using the Poisson summation formula [34], evaluation of trigonometric integrals by the modulus of derivatives [35] and evaluation of full rational sums by Hua Lo-ken [36].

In the third paragraph of the first chapter, a nontrivial evaluation of short trigonometric Weyl sums in small arcs is found in small arcs $T(\alpha; x, y)$ of degree four.

Theorem 1.3. *Let* $x \geq x_0 > 0$, $y_0 < y \leq 0{,}01x$, α - *is a real number,*

$$\left|\alpha - \frac{a}{q}\right| \leq \frac{1}{q^2}, \qquad (a, q) = 1.$$

Then it is fair to estimate

$$|T(\alpha; x, y)| \ll y\left(q^{-\frac{1}{16}} + y^{-\frac{1}{16}}\ln^{\frac{1}{16}}q + y^{-\frac{1}{4}}q^{\frac{1}{16}}\ln^{\frac{1}{16}}q\right)(\ln y)^{\frac{7}{16}}.$$

The proof of Theorem 1.3 relies on the following lemma, which in turn is proved by the method of G. Weyl.

Lemma 1.3. *Let* x *и* y - *are real numbers,* $1 \leq y < x$,

$$T(\alpha; x, y) = \sum_{x-y<m<x} e(\alpha m^4).$$

Then the relation

$$|T(\alpha; x, y)|^8 \leq$$

$$2^{11}y^4 \sum_{0<k\leq y} \sum_{0<r\leq y-k} \sum_{0<t<y-k-r} \left|\sum_{x-y<m<x-k-r-t} e(24\alpha krtm)\right| + 2^{11}y^7.$$

Esterman [37] proved an asymptotic formula for the number of solutions of Eq.

$$p_1 + p_2 + m^2 = N, \quad (3)$$

where p_1, p_2 - are prime numbers, m - is a natural number. As we have already mentioned in [18, 23], this problem is investigated with more rigid conditions, namely, when the summands are almost equal, and we derive an asymptotic formula for the number of solutions of (1.1) with the following conditions

$$\left|p_i - \frac{N}{3}\right| \leq H; \qquad i = 1,2, \qquad \left|m^2 - \frac{N}{3}\right| \leq H; \qquad H \geq$$

$N^{\frac{3}{4}}\ln^3 N.$

Further, in [20, 24] the asymptotic formula is derived for a rarer sequence with almost equal summands, i.e., when in equation (1.1) the square of natural m is replaced by its cube at $H \geq N^{\frac{5}{6}}\mathcal{L}^{10}$. In the second chapter, adjoining the results of the previous chapters, viz:

- Theorem 2.1 on the behavior of short trigonometric sums of G. Weyl $T(\alpha; x, y)$ in large arcs;

- Theorem 2.3. on non-trivial evaluation of short trigonometric sums of G. Weyl $T(\alpha; x, y)$ of degree four in small arcs,

we prove Theorem 1 on the asymptotic formula for an even rarer sequence with almost equal summands, i.e., when in Eq. (1.1) the square of natural m is replaced by the fourth degree.

Theorem 2.1. *Let N - a sufficiently large natural number, $I(N, H)$ - number of representations N by the sum of two prime numbers p_1, p_2 and the fourth degree of the natural m under the conditions*

$$\left|p_i - \frac{N}{3}\right| \leq H, \qquad i = 1,2, \qquad \left|m^4 - \frac{N}{3}\right| \leq H,$$

$\rho(N,p)$ - number of comparison solutions $x^4 \equiv N(\quad \mathrm{mod}\,p)$. Then at $H \geq N^{\frac{11}{12}}\mathcal{L}^{\frac{40}{3}}$ the asymptotic formula is valid:

$$I(N,H) = \frac{\sqrt[4]{3}\,\mathfrak{S}(N)\;H}{4\sqrt[4]{N^3}\,\mathcal{L}^2} + O\left(\frac{H^2}{\sqrt[4]{N^3}\,\mathcal{L}^3}\right), \qquad \mathfrak{S} = \prod_p \left(1 + \frac{\rho(N,p)}{(p-1)^2}\right).$$

Corollary 2.1. *There exists such N_0 that every natural number $N > N_0$ is representable as a sum of two prime numbers p_1, p_2 and the fourth degree of the natural number m with the conditions*

$$\left|p_i - \frac{N}{3}\right| \leq N^{\frac{11}{12}}\mathcal{L}^{\frac{40}{3}}, \qquad i = 1,2,$$

$$\left|m - \sqrt[4]{\frac{N}{3}}\right| \leq \frac{3N^{\frac{1}{6}}\mathcal{L}^{\frac{40}{3}}}{4\sqrt[4]{3}} + \frac{27N^{\frac{1}{12}}\mathcal{L}^{\frac{80}{3}}}{32\sqrt[4]{3}} + \frac{189\mathcal{L}^{40}}{128\sqrt[4]{3}} + 0{,}9.$$

The proof of Theorem 1 is carried out by the circular method of Hardy, Littlewood, Ramanujan in the form of trigonometric sums of I.M. Vinogradov. Its basis, as already mentioned, is the corollaries 2, 2 of Theorem 2.1. and Theorem 2.3.

Chapter 1. Short Trigonometric Sums by G. Weyl

1.1. Auxiliary lemmas

Let's say H и y are arbitrary integers, $H \geq 1$. Then the relation

$$\sum_{x=y+1}^{y+H} e(\alpha x) \leq \min\left(H, \frac{1}{2\|\alpha\|}\right), \qquad \| \alpha \| = \min\{\alpha, 1 - \alpha\}.$$

For the proof see. [38].

Let's

$$\alpha = \frac{a}{q} + \frac{\theta}{q^2}, \qquad (a, q) = 1, \qquad q \geq 1, \qquad |\theta| \leq 1.$$

Then at β, $U > 0$, $P > 1$ we have

$$\sum_{x=1}^{P} \min\left(U, \frac{1}{\|\alpha x + \beta\|}\right) \leq 6\left(\frac{P}{q} + 1\right)(U + q\ln q).$$

For the proof see. [38].

Given $x \geq 2$ we have

$$\sum_{n \leq x} \tau_r^2(n) \ll x(\ln x)^{r^2 - 1}$$

For the proof see. [39]

Let $f(u)$ - a real function, $f''(u) > 0$ in the interval $[a, b]$, α, β, ε arbitrary numbers with the conditions $\alpha \leq f'(a) \leq f'(b) \leq \beta$ и $0 < \varepsilon \leq 1$. Then

$$\sum_{a < n \leq b} e(f(n)) = \sum_{\alpha - \varepsilon < h \leq \beta + \varepsilon} \int_a^b e(f(u) - hu)du + O(\varepsilon^{-1} +$$

$\ln(\beta - \alpha + 2))$,

where the constant in the sign O is absolute.

For the proof see. [34].

Let $(a, q) = 1$, q - is a natural number, b - is an arbitrary integer. Then we have

$$S_b(a, q) = \sum_{k=1}^{q} e\left(\frac{ak^n + bk}{q}\right) \ll q^{1/2+\varepsilon}(b, q).$$

For the proof see. [36].

Let a real function $f(u)$ and a monotone function $g(u)$ satisfy the conditions: $f'(u)$ - monotone, $|f'(u)| \geq m > 0$ и $|g(u)| \leq M$. Then the estimation is valid:

$$\int_a^b g(u)e(f(u))du \ll \frac{M}{m}.$$

For the proof see. [1].

Let at $a \leq u \leq b$ real function $f(u)$ has a derivative n - of order $(n > 1)$, and at some $A > 0$ the inequality $A \leq |f^{(n)}(u)|$. Then the estimate

$$\int_a^b e(f(u))du \leq \min(b - a, 6nA^{-\frac{1}{n}}).$$

For the proof see. [35].

Let $n \geq 3$ - integer and $f(t) = a_n t^n + a_{n-1} t^{n-1} + \cdots + a_1 t$ - a polynomial with integer coefficients, $(a_n, \ldots, a_1, q) = 1$, q - is a natural number. Then we have

$$|S(q, f(t))| = \left|\sum_{k=1}^{q} e\left(\frac{f(k)}{q}\right)\right| \leq c(n)q^{1-\frac{1}{n}},$$

where

$$c(n) = \begin{cases} \exp(4n), & \text{при} \quad n \geq \quad 10; \\ \exp(n(A(n))), & \text{при} \quad 3 \leq \quad n \leq \quad 9. \end{cases}$$

$$A(3) = 6{,}1, \qquad A(4) = 5{,}5, \qquad A(5) = 5, \qquad A(6) = 4{,}7,$$

$$A(7) = 4{,}4, \qquad A(8) = 4.2, \qquad A(9) = 4{,}05.$$

For the proof see. [35].

1.2. Behavior of short trigonometric sums of G. Weyl in large arcs.

P. Vaughn [31] studying G. Weyl sums of the form

$$T(\alpha, x) = \sum_{m \leq x} e(\alpha m^n), \qquad \alpha = \frac{a}{q} + \lambda, \qquad q \leq$$

τ, $\quad (a, q) = 1, \quad |\lambda| \leq \frac{1}{q\tau}$,

in large arcs using the estimate

$$S_b(a, q) = \sum_{k=1}^{q} e\left(\frac{ak^n + bk}{q}\right) \ll q^{\frac{1}{2} + \varepsilon}(b, q), \qquad (2.1)$$

belonging to Hua Lo-ken [36], by van der Korput's method proved:

$$T(\alpha, x) = \frac{S(a,q)}{q} \int_0^x e(\lambda t^n) dt + O\left(q^{\frac{1}{2} + \varepsilon}(1 + $$

$x^n |\lambda|)^{\frac{1}{2}}\Big).$ $\qquad S(a, q) = S_0(a, q),$

Provided that α is very well approximated by a rational number with denominator q that is, if the condition

$$|\lambda| \leq \frac{1}{2nqx^{n-1}},$$

he also proved

$$T(\alpha, x) = \frac{x}{q} \cdot \frac{S(a,q)}{q} \int_0^1 e(\lambda t^n) dt + O\left(q^{\frac{1}{2} + \varepsilon}\right).$$

He used these estimates to derive an asymptotic formula in the Waring problem for eight cubes [40].

Short trigonometric sums of G. Weyl of the form

$$T(\alpha; x, y) = \sum_{x - y < m \leq x} e(\alpha m^n), \qquad \alpha = \frac{a}{q} + \lambda, \ q \leq$$

τ, $\quad (a, q) = 1, |\lambda| \leq \frac{1}{q\tau}$, $\qquad\qquad (2.2)$

derived from $T(\alpha, x)$ by replacing the condition $m \leq x$ by the condition $x - y < m \leq x$, in large arcs at $n = 2,3,4$ were studied in [18, 19, 21, 20, 22] and applied in the derivation of asymptotic formulas with almost equal summands in the Waring problem (for cubes and fourth powers) in [25, 26] and the cubic Esterman problem in [24, 20]. Then for arbitrary fixed n sum $T(\alpha; x, y)$ was studied in [32, 33]. The main result of this work is a simplification of the proof and clarification of the main theorem of [32, 33].

Let's say $\tau \geq 2n(n-1)x^{n-2}y$ и $\lambda \geq 0$, then at $\{n\lambda x^{n-1}\} \leq \frac{1}{2q}$ the formula

$$T(\alpha, x, y) = \frac{S(a,q)}{q} T(\lambda; x, y) + O\left(q^{\frac{1}{2}+\varepsilon}\right),$$

and at $\{n\lambda x^{n-1}\} > \frac{1}{2q}$ there is an estimation

$$|T(\alpha, x, y)| \ll q^{1-\frac{1}{n}}\ln q + \min_{2 \leq k \leq n} \left(yq^{-\frac{1}{n}}, \lambda^{-\frac{1}{k}}x^{1-\frac{n}{k}}q^{-\frac{1}{n}}\right).$$

Let $\tau \geq 2n(n-1)x^{n-2}y$, $|\lambda| \leq \frac{1}{2nqx^{n-1}}$, then the relation

$$T(\alpha, x, y) = \frac{y}{q}S(a,q)\gamma(\lambda; x, y) + O\left(q^{\frac{1}{2}+\varepsilon}\right),$$

$$\gamma(\lambda; x, y) = \int_{-0,5}^{0,5} e\left(\lambda\left(x - \frac{y}{2} + yt\right)^n\right) dt.$$

Let $\tau \geq 2n(n-1)x^{n-2}y$, $\frac{1}{2nqx^{n-1}} < |\lambda| \leq \frac{1}{q\tau}$, then the estimate

$$T(\alpha, x, y) \ll q^{1-\frac{1}{n}}\ln q + \min_{2 \leq k \leq n}\left(yq^{-\frac{1}{n}}, x^{1-\frac{1}{k}}q^{\frac{1}{k}-\frac{1}{n}}\right).$$

Corollaries 2.1. and 2.2. are generalizations of results of R. Vaughn [31] for short trigonometric sums of G. Weyl $T(\alpha; x, y)$ of the form (2.2.).

The proof of Theorem 2.1. is carried out by van der Korput's method of evaluating special trigonometric sums using the Poisson summation formula [34], the evaluation of trigonometric integrals by the magnitude of the modulus of derivatives [35] and the evaluation of full rational sums (2.2.) belonging to Hua Lo-ken [36].

Proof of Theorem 2.1. Using the orthogonal property of the full linear

18

rational trigonometric sum, we find

$$T(\alpha; x, y) = \sum_{x-y<m\leq x} e\left(\frac{ak^n}{q} + \lambda m^n\right) \sum_{\substack{k=1 \\ k\equiv m(mod q)}}^{q} 1 =$$

$$= \sum_{k=1}^{q} e\left(\frac{ak^n}{q}\right) \sum_{\substack{x-y<m\leq x \\ m\equiv k(mod q)}} e(\lambda m^n) =$$

$$= \sum_{k=1}^{q} e\left(\frac{ak^n}{q}\right) \sum_{x-y<m\leq x} e(\lambda m^n) \frac{1}{q}\sum_{b=1}^{q} e\left(\frac{b(k-m)}{q}\right) =$$

$$= \frac{1}{q}\sum_{b=1}^{q} T_b(\lambda; x, y)S_b(a, q), \tag{2.3}$$

where

$$T_b(\lambda; x, y) = \sum_{x-y<m\leq x} e\left(\lambda m^n - \frac{bm}{q}\right), \qquad T(\lambda; x, y) =$$

$T_0(\lambda; x, y),$

$$S_b(a, q) = \sum_{k=1}^{q} e\left(\frac{ak^n+bk}{q}\right), \qquad S(a, q) = S_0(a, q).$$

By $R(\alpha; x, y)$ denote the part of the sum $T(\alpha; x, y)$ which is defined by the relation (???), in which there is no summand at $b = q$ i.e.

$$R(\alpha; x, y) = \frac{1}{q}\sum_{b=1}^{q-1} T_b(\lambda; x, y)S_b(a, q). \tag{2.4}$$

Keeping in mind that $n\lambda x^{n-1} - \{n\lambda x^{n-1}\}$ - is an integer, let us represent $T_b(\lambda; x, y)$ in the form of

$$T_b(\lambda; x, y) = \sum_{x-y<m\leq x} e(f(m, b)),$$

$$f(u, b) = \lambda u^n - (n\lambda x^{n-1} - \{n\lambda x^{n-1}\})u - \frac{bu}{q}.$$

Find the first and second order derivative of the function $f(u, b)$:

$$f'(u, b) = n\lambda(u^{n-1} - x^{n-1}) + \{n\lambda x^{n-1}\} - \frac{b}{q},$$

$$f''(u, b) = n(n-1)\lambda u^{n-2} \geq 0.$$

Hence the function $f'(u, b)$, $u \in (x - y, x]$ is non-decreasing, so for all $u \in [x - y, x)$ and any b, $b = 1, 2, \ldots, q$ the inequality

$$f'(x - y, b) < f'(u, b) \leq f'(x, b). \qquad (2.5)$$

Estimating $f'(x, b)$ from above, we have:

$$f'(x, b) = \{n\lambda x^{n-1}\} - \frac{b}{q} < 1 - \frac{b}{q}, \qquad (2.6)$$

To estimate from below $f'(x - y, b)$ we use the representation

$$f'(x - y, b) = -n\lambda(x^{n-1} - (x - y)^{n-1}) + \{n\lambda x^{n-1}\} - \frac{b}{q} =$$

$$= n\lambda \sum_{k=1}^{n-1} (-1)^k C_{n-1}^k x^{n-1-k} y^k + \{n\lambda x^{n-1}\} - \frac{b}{q} =$$

$$= -n(n - 1)\lambda x^{n-2} y + n\lambda \sum_{k=2}^{n-1} (-1)^k C_{n-1}^k x^{n-1-k} y^k + \{n\lambda x^{n-1}\} - \frac{b}{q}.$$

Using the monotonicity $f'(u, b)$ condition $\tau \geq 2n(n - 1)x^{n-2}y$ and inequality

$$W = \sum_{k=2}^{n-1} (-1)^k C_{n-1}^k x^{n-1-k} y^k \geq 0, \qquad n \geq 3, \qquad 3x \geq (n -$$

$3)y$,

we have

$$f'(u, b) \leq f'(x, b) = \{n\lambda x^{n-1}\} - \frac{b}{q} < 1,$$

$$f'(u, b) \geq f'(x - y, b) = -n(n - 1)\lambda x^{n-2} y + n\lambda \ W + \{n\lambda x^{n-1}\} - \frac{b}{q} \geq$$

$$\geq -n(n - 1)\lambda x^{n-2} y - \frac{b}{q} \geq -\frac{n(n-1)x^{n-2}y}{q\tau} - \frac{b}{q} \geq -1 + \frac{1}{2q}.$$

Therefore, applying to the sum $T_b(\lambda; x, y)$ Poisson summation formula, (lemma 1) at $\alpha = -1$, $\beta = 1$, $\varepsilon = 0{,}5$ we obtain

$$T_b(\lambda; x, y) = I(-1, b) + I(0, b) + I(1, b) + O(1), \qquad (2.7)$$

$$I(h, b) = \int_{x-y}^{x} e(f_h(u, b))du, \qquad f_h(u, b) = f(u, b) - hu.$$

Function

$$f'_h(u, b) = n\lambda(u^{n-1} - x^{n-1}) + \{n\lambda x^{n-1}\} - \frac{b}{q} - h$$

on the interval $u \in [x - y, x]$ is a non-decreasing function, so

$$f'_h(x - y, b) \le f'_h(u, b) \le f'_h(x, b),$$

which can be represented as

$$\{n\lambda x^{n-1}\} - \frac{b}{q} - h - \eta < f'_h(u, b) \le \{n\lambda x^{n-1}\} - \frac{b}{q} - h, \quad (2.8)$$

$$\eta = n(n - 1)\lambda x^{n-2}y - n\lambda W \le n(n - 1)\lambda x^{n-2}y \le$$

$$\frac{n(n-1)x^{n-2}y}{q\tau} \le \frac{1}{2q}.$$

Then, substituting (???) into (???) and (???), we find

$$+T_0 + T_1 + O()T(\alpha; x, y) = T_{-1} + T_0 + T_1 +$$

$$O\left(\frac{1}{q}\sum_{b=0}^{q-1} |S_b(a, q)|\right), \qquad (2.9)$$

$$+R_0 + R_1 + O()R(\alpha; x, y) = R_{-1} + R_0 + R_1 +$$

$$O\left(\frac{1}{q}\sum_{b=1}^{q-1} |S_b(a, q)|\right), \qquad (2.10)$$

$$T_h = \frac{1}{q}\sum_{b=0}^{q-1} I(h, b)S_b(a, q),$$

$$R_h = \frac{1}{q}\sum_{b=1}^{q-1} I(h, b)S_b(a, q).$$

Using the estimate (???), we estimate the residual term:

$$\frac{1}{q}\sum_{b=1}^{q-1}|S_b(a,q)| \ll q^{-\frac{1}{2}+\varepsilon}\sum_{b=1}^{q-1}(b,q) =$$

$$q^{-\frac{1}{2}+\varepsilon}\sum_{\delta\backslash q}\delta\sum_{\substack{1\leq b\leq q-1\\(b,q)=\delta}} 1 \leq q^{\frac{1}{2}+\varepsilon}\tau(q).$$

Let's estimate each sum T_h и R_h separately.

Evaluation T_1 и R_1. Assuming $h = 1$ in (???), we have

$$f'_1(u,b) \leq \{n\lambda x^{n-1}\} - \frac{b}{q} - 1 \leq -\frac{b}{q} < 0.$$

Estimating the integral by the value of the first derivative (lemma 1), we have

$$|I(1,b)| = \left|\int_{x-y}^{x} e(f_1(u,b))du\right| \ll \frac{q}{b}.$$

Hence and from (???), we have

$$R_1 = \frac{1}{q}\sum_{b=1}^{q-1} I(1,b)S_b(a,q) \ll \sum_{b=1}^{q-1}\frac{|S_b(a,q)|}{b} \ll q^{\frac{1}{2}+\varepsilon}\sum_{b=1}^{q-1}\frac{(b,q)}{b} \ll q^{\frac{1}{2}+2\varepsilon}.$$

In the case of $b = 0$using the inequality

$$f_1^{(k)}(u,q) \geq n(n-1)\ldots(n-k+1)\lambda(x-y)^{n-k} \gg$$

$\lambda x^{n-k}, \qquad k = 2,3,\ldots,n,$

evaluating the integral $I(1,0)$ by the value of k - of the derivative (lemma 1), we find

$$|I(1,0)| \ll \min_{2\leq k\leq n}\left(y, \lambda^{-\frac{1}{k}}x^{1-\frac{n}{k}}\right).$$

Hence, using the estimation $|S(a,q)| \ll q^{1-\frac{1}{n}}$ (lemma 1), taking into account the estimation R_1 we obtain

$$T_1 \leq |R_1| + \frac{|I(1,0)||S(a,q)|}{q} \ll q^{\frac{1}{2}+2\varepsilon} +$$

$$\min_{2\leq k\leq n}\left(yq^{-\frac{1}{n}}, \lambda^{-\frac{1}{k}}x^{1-\frac{n}{k}}q^{-\frac{1}{n}}\right).$$

Evaluation T_{-1} и R_{-1}. Assuming $h = -1$ in (???), we have

$$f'_{-1}(u,b) > \{n\lambda x^{n-1}\} + \frac{q-b}{q} - \eta \geq \frac{q-b}{q}.$$

integral $I(-1, b)$ we also evaluate by the value of the first derivative (lemma 1). We have

$$|I(-1, b)| = \left| \int_{x-y}^{x} e(f_{-1}(u, b)) du \right| \ll \frac{q}{q-b}.$$

Hence, proceeding similarly to the case of estimation R_1 we obtain

$$R_{-1} = \sum_{b=1}^{q-1} \frac{I(-1,b)S_b(a,q)}{q} \ll \sum_{b=1}^{q-1} \frac{|S_b(a,q)|}{q-b} \ll$$

$$q^{\frac{1}{2}+\varepsilon} \sum_{b=1}^{q-1} \frac{(b,q)}{b} \ll q^{\frac{1}{2}+2\varepsilon}.$$

$$T_{-1} \leq |R_{-1}| + \frac{|I(-1,0)||S(a,q)|}{q} \ll q^{\frac{1}{2}+2\varepsilon} + \frac{|S_b(a,q)|}{q} \ll q^{\frac{1}{2}+2\varepsilon}.$$

Evaluation R_0. If $\{n\lambda x^{n-1}\} \leq \frac{1}{2q}$, then, assuming $h = 0$ in (???),

we have

$$f'_0(u, b) \leq \{n\lambda x^{n-1}\} - \frac{b}{q} \leq \frac{1-2b}{2q} \leq -\frac{b}{2q} < 0.$$

integral $I(0, b)$, also evaluating by the value of the first derivative, we find

$$|I(0, b)| = \left| \int_{x-y}^{x} e(f_0(u, b)) du \right| \ll \frac{q}{b}.$$

Proceeding similarly to the case of estimation R_1 we obtain

$$R_0 = \sum_{b=1}^{q-1} \frac{I(0,b)S_b(a,q)}{q} \ll \sum_{b=1}^{q-1} \frac{|S_b(a,q)|}{b} \ll q^{\frac{1}{2}+\varepsilon} \sum_{b=1}^{q-1} \frac{(b,q)}{b} \ll$$

$$q^{\frac{1}{2}+2\varepsilon}.$$

Hence, from the estimates R_1 и R_{-1} taking into account (???), we obtain the first statement of the theorem.

Evaluation T_0. Given $\{n\lambda x^{n-1}\} \geq \frac{1}{2q}$, we define the natural number r by the ratio

$$\frac{r}{2q} \leq \{n\lambda x^{n-1}\} < \frac{r+1}{2q}, \qquad 1 \leq r \leq 2q - 1.$$

Hence, from the inequality (???) at $h = 0$ and the condition $\eta \leq \frac{1}{2q}$ we find

$$f'_0(u,b) > \{n\lambda x^{n-1}\} - \frac{b}{q} - \eta \geq \frac{r-2b-1}{2q}, \qquad\qquad (2.11)$$

$$f'_0(u,b) \leq \{n\lambda x^{n-1}\} - \frac{b}{q} < \frac{r-2b+1}{2q}. \qquad\qquad (2.12)$$

Let $r = 2r_1$ - even $(1 \leq r_1 \leq q - 1)$. Summation segment $0 \leq b \leq q - 1$ in the sum T_0 let us split it into the following three sets:

$$0 \leq b \leq r_1 - 1, \qquad\qquad b = r_1, \qquad\qquad r_1 + 1 \leq b \leq q - 1,$$

respectively, in the first of which the right part of the inequality (???) is greater than zero, and in the third the right part of the inequality (???) is less than zero, i.e.

$$f'_0(u,b) > \frac{2r_1-2b-1}{2q} \geq \frac{r_1-b}{2q}, \qquad\qquad 0 \leq b \leq r_1 - 1,$$

$$f'_0(u,b) < \frac{2r_1-2b+1}{2q} \leq \frac{r_1-b}{2q}, \qquad\qquad r_1 + 1 \leq b \leq q - 1.$$

Using these inequalities, evaluating the integral $I(0,b)$ by the value of the first derivative, we find

$$I(0,b) = \int_{x-y}^{x} e(f_0(u,b))du \ll \frac{q}{|r_1-b|}, \qquad b \neq r_1.$$

In the case of $b = r_1$, evaluating similarly to the evaluation of the integral $I(1,0)$we find

$$|I(0,r_1)| \ll \min_{2 \leq k \leq n} \left(y, \lambda^{-\frac{1}{k}}x^{1-\frac{n}{k}}\right).$$

Using these estimates and the estimate of $|S(a,q)| \ll q^{1-\frac{1}{n}}$ (lemma 1), we obtain

$$T_0 = \sum_{b=0}^{q-1} \frac{I(0,b)S_b(a,q)}{q} \ll q^{-\frac{1}{n}}\left(\sum_{\substack{b=0, \\ b \neq r_1}}^{q-1} \frac{q}{|r_1-b|} + \right.$$

$$\left. \min_{2 \leq k \leq n} \left(y, \lambda^{-\frac{1}{k}}x^{1-\frac{n}{k}}\right)\right) \ll$$

$$\ll q^{1-\frac{1}{n}}\ln q + \min_{2 \leq k \leq n} \left(yq^{-\frac{1}{n}}, \lambda^{-\frac{1}{k}}x^{1-\frac{n}{k}}q^{-\frac{1}{n}}\right).$$

Let now $r = 2r_1 + 1$ - odd $(0 \leq r_1 \leq q - 1)$. The summation segment

$0 \leq b \leq q - 1$ in the sum R_0 let us split it into the following three sets:

$$0 \leq b \leq r_1 - 1, \qquad b = r_1, r_1 + 1, \qquad r_1 + 2 \leq b \leq q - 1,$$

respectively, in the first of which the right part of the inequality (???) is greater than zero, and in the third - the right part of the inequality (???) is less than zero, i.e.

$$f'_0(u, b) > \frac{2r_1 + 1 - 2b - 1}{2q} = \frac{r_1 - b}{q}, \qquad 0 \leq b \leq r_1 - 1,$$

$$f'_0(u, b) < \frac{2r_1 + 1 - 2b + 1}{2q} \leq \frac{r_1 - b}{2q}, \qquad r_1 + 2 \leq b \leq q -$$

1.

Hence,

$$I(0, b) = \int_{x-y}^{x} e(f_0(u, b)) du \ll \frac{q}{|r_1 - b|}, \qquad b \neq r_1 - 1, r_1.$$

In the case of $b = r_1 - 1, r_1$, proceeding similarly to the previous estimation $I(0, r_1)$ find

$$|I(0, b)| \ll \min_{2 \leq k \leq n} \left(y, \lambda^{-\frac{1}{k}} x^{1 - \frac{n}{k}} \right), \qquad b = r_1, \; r_1 + 1.$$

From these estimates for $I(0, b)$ we obtain

$$T_0 \leq \frac{1}{q} \sum_{b=0}^{q-1} |I(0, b)| |S_b(a, q)| \ll$$

$$\ll q^{-\frac{1}{n}} \left(\sum_{\substack{b=0, \\ b \neq r_1, r_1 + 1}}^{q-1} \frac{q}{|r_1 - b|} + \min_{2 \leq k \leq n} \left(y, \lambda^{-\frac{1}{k}} x^{1 - \frac{n}{k}} \right) \right) \ll$$

$$\ll q^{1 - \frac{1}{n}} \ln q + \min_{2 \leq k \leq n} \left(y q^{-\frac{1}{n}}, \lambda^{-\frac{1}{k}} x^{1 - \frac{n}{k}} q^{-\frac{1}{n}} \right).$$

Substituting the estimates for T_1, T_{-1} и T_0 in (???), we obtain the second statement of the theorem.

Note. The case $\lambda < 0$ reduces to the case $\lambda \geq 0$ if we give the formula (???) the form

$$\overline{T(\alpha; x, y)} = \frac{1}{q} \sum_{b=0}^{q-1} T_{q-b}(-\lambda; x, y) S_{q-b}(q - a, q) =$$

$$\frac{1}{q} \sum_{b=0}^{q-1} T_b(-\lambda; x, y) S_b(q - a, q).$$

1.3. Estimation of G. Weyl's short trigonometric sums of fourth order in small arcs

When deriving asymptotic formulas in additive problems with almost equal summands, which include the Varing problem and the Esterman problem, the main point, along with the circular Hardy-Littlewood method in the variant of trigonometric sums of I. M. Vinogradov, is also the behavior of short trigonometric sums of G. Weyl of the form

$$T(\alpha; x, y) = \sum_{x-y<m\leq x} e(\alpha m^n), \quad \alpha = \frac{a}{q} + \lambda, \qquad (a, q) =$$

$$1, \qquad q \leq \tau, \ |\lambda| \leq \frac{1}{q\tau},$$

in large arcs and their evaluation in small arcs. The behavior $T(\alpha; x, y)$ in large arcs was studied in the previous paragraph. In this paragraph, using the method of G. Weyl, we will find a nontrivial estimate of the short Weyl trigonometric sum of fourth order.

Let's say x и y - are real numbers, $1 \leq y < x$,

$$T(\alpha; x, y) = \sum_{x-y<m<x} e(\alpha m^4).$$

Then the relation

$$|T(\alpha; x, y)|^8 \leq$$

$$2^{11}y^4 \sum_{0<k\leq y} \sum_{0<r\leq y-k} \sum_{0<t<y-k-r} \left|\sum_{x-y<m<x-k-r-t} e(24\alpha krtm)\right| + 2^{11}y^7.$$

Proof. Let us transform $|T(\alpha; x, y)|^2$. We have

$$|T(\alpha; x, y)|^2 = \sum_{x-y<m,n\leq x} e(\alpha(m^4 - n^4)) =$$

$$= \sum_{x-y<m<n\leq x} e(\alpha(m^4 - n^4)) + \sum_{x-y<n<m\leq x} e(\alpha(m^4 - n^4)) + \sum_{x-y<m\leq x} 1 \leq$$

$$\leq 2\left|\sum_{x-y<m<x} \sum_{m<n\leq x} e(\alpha(n^4 - m^4))\right| + y =$$

$$= 2\left|\Sigma_{x-y<m<x} \quad \Sigma_{0<k\leq x-m} \; e(\alpha((m+k)^4 - m^4))\right| + y.$$

Hence, using the identity

$$(m+k)^4 - m^4 = kf_1(m) + k^4, \qquad\qquad f_1(m) = 4m^3 +$$

$6km^2 + 4k^2 m,$

let's find

$$|T(\alpha; x, y)|^2 \leq$$

$$2\left|\Sigma_{0<k\leq y} \; e(\alpha k^4) \Sigma_{x-y<m\leq x-k} \; e(\alpha k f_1(m))\right| + y \leq$$

$$\leq 2\Sigma_{0<k\leq y} \; |W(k)| + y, \qquad W(k) =$$

$\Sigma_{x-y<m\leq x-k} \; e(\alpha k f_1(m)).$

Elevating both parts of the inequality to the fourth degree, then using twice successively the ratio $(a+b)^2 \leq 2a^2 + 2b^2$ and Cauchy inequality, we find

$$|T(\alpha; x, y)|^8 \leq \left(2\Sigma_{0<k\leq y} \; |W(k)| + y\right)^4 \leq$$

$$\left(2^3\left(\Sigma_{0<k\leq y} \; |W(k)|\right)^2 + 2y^2\right)^2 \leq$$

$$\leq \left(2^3 y \Sigma_{0<k\leq y} \; |W(k)|^2 + 2y^2\right)^2 \leq$$

$$2^7 y^2 \left(\Sigma_{0<k\leq y} \; |W(k)|^2\right)^2 + 2^3 y^4 \leq$$

$$\leq 2^7 y^3 \Sigma_{0<k\leq y} \; |W(k)|^4 + 2^3 y^4. \qquad\qquad (3.1)$$

Let's convert $|W(k)|^2$. We have

$$|W(k)|^2 = \Sigma_{x-y<m,n\leq x-k} \; e(\alpha k(f_1(n) - f_1(m))) \leq$$

$$\leq 2\left|\Sigma_{x-y<m\leq x-k} \quad \Sigma_{m<n\leq x-k} \; e(\alpha k(f_1(n) - f_1(m)))\right| +$$

$y =$

$$\leq 2\left|\Sigma_{x-y<m<x-k} \quad \Sigma_{0<r\leq x-k-m} \; e(\alpha k(f_1(m+r) -$$

$f_1(m)))| + y =$

$$= 2\left|\Sigma_{0<r\leq y-k} \quad \Sigma_{x-y<m<x-k-r} \; e(\alpha k(f_1(m+r) -$$

$f_1(m)))| + y.$

Here using the identity

$$f_1(m+r) - f_1(m) = 4((m+r)^3 - m^3) + 6k((m+r)^2 - m^2) + 4k^2 r =$$

$$= 4(3m^2 r + 3mr^2 + r^3) + 6k(2mr + r^2) + 4k^2 r =$$

$$= 12m^2 r + 12m(kr + r^2) + 4k^2 r + 6kr^2 + 4r^3 =$$

$$= 12r f_2(m) + 4k^2 r + 6kr^2 + 4r^3,$$

$$f_2(m) = m^2 + m(k+r),$$

we find

$$|W(k)|^2 \le 2 \Big| \sum_{0 < r \le y-k} e(\alpha k(4k^2 r + 6kr^2 + 4r^3)) \sum_{x-y < m < x-k-r} e(12\alpha k r f_2(m)) \Big| + y \le$$

$$\le 2 \sum_{0 < r \le y-k} |W(k,r)| + y, \qquad (3.2)$$

$$W(k,r) =$$

$$\sum_{x-y<m<x-k-r} e(12\alpha k r f_2(m)).$$

Then using the identity

$$f_2(m+t) - f_2(m) = (m+t)^2 + (m+t)(k+r) - m^2 - m(k+r) =$$

$$= 2mt + t^2 + (k+r)t,$$

and proceeding similarly as in the case $W(k)$ we find:

$$|W(k,r)|^2 \le$$

$$2 \Big| \sum_{x-y<m<x-k-r} \sum_{m<n<x-k-r} e(12\alpha k r(f_2(n) - f_2(m))) \Big| + y =$$

$$= 2 \Big| \sum_{x-y<m<x-k-r} \sum_{0<t<x-k-r-m} e(12\alpha k r(f_2(m+t) - f_2(m))) \Big| + y =$$

$$= 2 \Big| \sum_{0<t<y-k-r} \sum_{x-y<m<x-k-r-t} e(12\alpha k r(f_2(m+t) - f_2(m))) \Big| + y =$$

$$= 2 \Big| \sum_{0<t<y-k-r} e(12\alpha k r(t^2 + (k+r)t)) \sum_{x-y<m<x-k-r-t} e(24\alpha k r t m) \Big| + y =$$

$$= 2 \sum_{0<t<y-k-r} \Big| \sum_{x-y<m<x-k-r-t} e(24\alpha k r t m) \Big| + y. \qquad (3.3)$$

Successively substituting into (???) the values of $W(k)$ и $W(k,r)$ respectively from (???) and (???), each time using the ratio $(a+b)^2 \leq 2a^2 + 2b^2$ and Cauchy's inequality, we find

$$|T(\alpha; x, y)|^8 \leq 2^7 y^3 \sum_{0<k\leq y} |W(k)|^4 + 2^3 y^4 \leq$$

$$\leq 2^7 y^3 \sum_{0<k\leq y} \left(2 \sum_{0<r\leq y-k} |W(k,r)| + y\right)^2 + 2^3 y^4 \leq$$

$$\leq 2^7 y^3 \sum_{0<k\leq y} \left(2^3 \left(\sum_{0<r\leq y-k} |W(k,r)|\right)^2 + 2y^2\right) +$$

$$2^3 y^4 \leq$$

$$\leq 2^{10} y^4 \sum_{0<k\leq y} \sum_{0<r\leq y-k} |W(k,r)|^2 + 2^8 y^6 + 2^3 y^4 \leq$$

$$\leq$$

$$2^{10} y^4 \sum_{0<k\leq y} \sum_{0<r\leq y-k} \left(2 \qquad \sum_{0<t<y-k-r} \left|\sum_{x-y<m<x-k-r-t} \qquad e(24\alpha krtm)\right| + y\right) + 2^9 y^6 \leq$$

$$\leq$$

$$2^{11} y^4 \sum_{0<k\leq y} \sum_{0<r\leq y-k} \sum_{0<t<y-k-r} \left|\sum_{x-y<m<x-k-r-t} \qquad e(24\alpha krtm)\right| + 2^{11} y^7.$$

Let $x \geq x_0 > 0$, $y_0 < y \leq 0{,}01x$, α - is a real number,

$$\left|\alpha - \frac{a}{q}\right| \leq \frac{1}{q^2}, \qquad (a, q) = 1.$$

Then it is fair to estimate

$$|T(\alpha; x, y)| \ll y \left(q^{-\frac{1}{16}} + y^{-\frac{1}{16}}\ln^{\frac{1}{16}}q + y^{-\frac{1}{4}}q^{\frac{1}{16}}\ln^{\frac{1}{16}}q\right) (\ln y)^{\frac{7}{16}}.$$

Proof. Applying to the sum by m in lemma 3, then lemma 1, we have

$$|T(\alpha; x, y)|^8 \leq$$

$$2^{11} y^4 \sum_{0<k\leq y} \sum_{0<r\leq y-k} \sum_{0<t<y-k-r} \left|\sum_{x-y<m<x-k-r-t} \qquad e(24\alpha krtm)\right| + 2^{11} y^7 \ll$$

$$\leq 2^{11} y^4 \sum_{0<k\leq y} \sum_{0<r\leq y-k} \sum_{0<t<y-k-r} \min\left(y, \frac{1}{\|24\alpha krt\|}\right) +$$

$$2^{11} y^7 =$$

$$= 2^{11}y^4 \sum_{n \le 24y^3} \eta(n)\min\left(y, \frac{1}{\|\alpha n\|}\right) + 2^{11}y^7,$$

$$\eta(n) = \sum_{0<k\le y} \sum_{0<r\le y-k} \sum_{\substack{0<t<y-k-r \\ n=24krt}} 1 \le$$

$\tau_3(n),$

Applying Cauchy's inequality, lemma 1, then lemma 1, we obtain

$$|T(\alpha; x, y)|^{16} \ll$$

$$y^8 \sum_{n \le 24y^3} |\eta(n)|^2 \sum_{n \le 24y^3} \min\left(y, \frac{1}{\|\alpha n\|}\right)^2 + y^{14} \ll$$

$$\ll y^9 \sum_{n \le 24y^3} \tau_3^2(n) \sum_{n \le 24y^3} \min\left(y, \frac{1}{\|\alpha n\|}\right) + y^{14} \ll$$

$$\ll y^{12}\ln^7 y \sum_{n \le 24y^3} \min\left(y, \frac{1}{\|\alpha n\|}\right) + y^{14} \ll$$

$$\ll y^{12}\ln^7 y \left(\frac{y^3}{q} + 1\right)(y + q\ln q) + y^{14} =$$

$$= y^{16}\left(\frac{1}{q} + \frac{\ln q}{y} + \frac{1}{y^3} + \frac{q\ln q}{y^4}\right)\ln^7 y + y^{14}$$

$$\ll y^{16}\left(\frac{1}{q} + \frac{\ln q}{y} + \frac{q\ln q}{y^4}\right)\ln^7 y.$$

Hence,

$$|T(\alpha; x, y)| \ll y\left(q^{-\frac{1}{16}} + y^{-\frac{1}{16}}\ln^{\frac{1}{16}} q + y^{-\frac{1}{4}}q^{\frac{1}{16}}\ln^{\frac{1}{16}} q\right)(\ln y)^{\frac{7}{16}}.$$

The theorem is proved.

<u>**Chapter 2. Asymptotic formula in the Esterman problem of**</u>
<u>**degree four with almost equal summands.**</u>

2.1 Results formulation

T. Esterman [37] proved an asymptotic formula for the number of solutions of Eq.

$$p_1 + p_2 + m^2 = N, \qquad (1.1)$$

where p_1, p_2 - are prime numbers, m - natural number.

In [18] and [23] this problem is investigated with more rigid conditions, namely, when the summands are almost equal, and an asymptotic formula for the number of solutions of (1.1) with the conditions of

$$\left| p_i - \frac{N}{3} \right| \leq H; \quad i = 1,2, \qquad \left| m^2 - \frac{N}{3} \right| \leq H; \quad H \geq N^{\frac{3}{4}} \ln^3 N.$$

Further in [20] the asymptotic formula is derived for a rarer sequence with almost equal summands, i.e., when in equation (1.1) the square of natural m is replaced by its cube at $H \geq N^{\frac{5}{6}} \mathcal{L}^{10}$. (see also [24]).

The main result of the third chapter is Theorem 1 on the asymptotic formula for an even rarer sequence with almost equal summands when in equation (1.1) the square of the natural m is replaced by the fourth degree. Theorem 1 is proved by the circular method of Hardy, Littlewood, and Ramanujan in the form of trigonometric sums by I. M. Vinogradov, using the results of the previous chapters, viz.

- Theorem 2 on the behavior of short trigonometric sums of G. Weyl $T(\alpha; x, y)$ for α belonging to large arcs;

- Theorem 3 on non-trivial evaluation of short trigonometric sums of G. Weyl $T(\alpha; x, y)$ of degree four in small arcs;

- Lemma 2 on the behavior of short linear trigonometric sums with prime numbers $S(\alpha; x, y)$ for α, belonging to large arcs.

Let N - a sufficiently large natural number, $I(N, H)$ - number of representations N by the sum of two prime numbers p_1, p_2 and the fourth degree of the natural m under the conditions

$$\left| p_i - \frac{N}{3} \right| \leq H, \qquad i = 1,2, \qquad \left| m^4 - \frac{N}{3} \right| \leq H,$$

$\rho(N, p)$ - number of comparison solutions $x^4 \equiv N(\quad \mathrm{mod}p)$. Then at $H \geq N^{\frac{11}{12}}\mathcal{L}^{\frac{40}{3}}$ the asymptotic formula is valid:

$$I(N, H) = \frac{\sqrt[4]{3}\mathfrak{S}(N)\ H}{4\sqrt[4]{N^3}\mathcal{L}^2} + O\left(\frac{H^2}{\sqrt[4]{N^3}\mathcal{L}^3}\right), \qquad \mathfrak{S} = \prod_p \left(1 + \frac{\rho(N,p)}{(p-1)^2}\right).$$

There exists such a thing N_0 that every natural number $N > N_0$ is representable as the sum of two prime numbers p_1, p_2 and the fourth degree of a natural number m with the conditions

$$\left| p_i - \frac{N}{3} \right| \leq N^{\frac{11}{12}}\mathcal{L}^{\frac{40}{3}}, \qquad i = 1,2,$$

$$\left| m - \sqrt[4]{\frac{N}{3}} \right| \leq \frac{3N^{\frac{1}{6}}\mathcal{L}^{\frac{40}{3}}}{4\sqrt[4]{3}} + \frac{27N^{\frac{1}{12}}\mathcal{L}^{\frac{80}{3}}}{32\sqrt[4]{3}} + \frac{189\mathcal{L}^{40}}{128\sqrt[4]{3}} + 0,9.$$

2.2. Supporting Statements

Let p - is a prime number, $(a, p) = 1$, then the estimate

$$|S(a, p)| \ll \sqrt{p}.$$

For the proof, see[41].

Let $y \geq x^{\frac{7}{12}+\varepsilon}$, then the asymptotic formula is valid

$$\pi(x) - \pi(x-y) = \frac{y}{\ln x} + O\left(\frac{y}{\ln^2 x}\right).$$

For the proof see. [42].

Let $x \geq x_0$, $h \leq x^{\frac{3}{16}}\exp(-(\ln x)^{0,76})$,, $y \geq hx^{\frac{5}{8}}\exp(\ln x)^{0,76}$, $\tau \geq y^2/xh$, $q \leq h$, $b \geq 224$ - is an arbitrary fixed positive number,

$$F(q,x) = \begin{cases} \exp(-\ln^4 \ln x) & \text{еслиq} \leq (\ln x)^b, \\ (\ln x)^{b+3} & \text{еслиq} > (\ln x)^b. \end{cases}$$

Then the equality is true:

$$S(\alpha; x, y) = \sum_{x-y < n \leq x} \Lambda(n)e(\alpha n) = \frac{\mu(q)}{\varphi(q)} \frac{\sin \pi \lambda y}{\pi \lambda} e\left(\lambda\left(x - \frac{y}{2}\right)\right) +$$

$$O\left(\frac{y}{q^{1/2}} F(q,x)\right).$$

For the proof, see[43].

Let $q = q_1 q_2$, $(q_1, q_2) = 1$, $(a, q_1 q_2) = 1$then there exist singular numbers a_1 и $a_2$such that

$$a = a_1 q_2 + a_2 q_1, \qquad (a_1, q_1) = 1, \qquad 1 \leq a_1 \leq$$
q_1, $\qquad (a_2, q_2) = 1$, $\qquad 1 \leq a_2 \leq q_2$,
for which the relation

$$S(a, q) = S(a_1, q_1)S(a_2, q_2).$$

Proof. Since in the sum $S(a, q)$ any deduction x modulo $q = q_1 q_2$ can be represented in a unique way in the form

$$x \equiv x_1 q_2 + x_2 q_1 \qquad (\text{mod} q_1 q_2), \qquad\qquad 1 \leq x_1 \leq$$
q_1, $\qquad 1 \leq x_2 \leq q_2$,
we have

$$S(a, q) = S(a_1 q_2 + a_2 q_1, q_1 q_2) =$$

$$= \sum_{x_1=1}^{q_1} \sum_{x_2=1}^{q_2} e\left(\frac{(a_1 q_2 + a_2 q_1)(x_1 q_2 + x_2 q_1)^4}{q_1 q_2}\right) =$$

$$= \sum_{x_1=1}^{q_1} e\left(\frac{a_1(x_1 q_2)^4}{q_1}\right) \sum_{x_2=1}^{q_2} e\left(\frac{a_2(x_2 q_1)^4}{q_2}\right) =$$

$$= \sum_{x_1=1}^{q_1} e\left(\frac{a_1 x_1^4}{q_1}\right) \sum_{x_2=1}^{q_2} e\left(\frac{a_2 x_2^4}{q_2}\right) = S(a_1, q_1) S(a_2, q_2).$$

Let q - the number free of squares, $(a, q) = 1$, then the estimate

$$|S(a, q)| \ll \sqrt{q}.$$

Proof. Let $q = p_1 p_2 \dots p_k$ - the canonical decomposition of a number q into prime factors. According to lemma 2, there exist singular numbers $a_1, a_2 \dots, a_k$ such that

$$a = a_1 q_1 + a_2 q_2 + \cdots + a_k q_k, \quad q_i = q p_i^{-1}, \quad (a_i, p_i) = 1, \quad 1 \le a_i \le$$
$p_i, \ i = 1,2, \dots, k,$

for which the relation

$$S(a, q) = S(a_1, p_1) S(a_2, p_2) \dots S(a_k, p_k).$$

Hence, using A. Weyl's estimate (lemma 2), we obtain the statement of the lemma.

2.3. Proof of Theorem 1

Without limiting generality, we will assume that $H = N^{\frac{11}{12}} \mathcal{L}^{\frac{40}{3}}$. Let

$$\tau = 16 H N^{-\frac{1}{4}}, \qquad \ae\tau = 1.$$

We have

$$I(N, H) = \int_{-\ae}^{1-\ae} S_1^2(\alpha; N, H) T_1(\alpha; N, H) e(-\alpha N) d\alpha,$$

$$S_1(\alpha; N, H) = \sum_{|p - N/3| \le H} e(\alpha p);$$

$$T_1(\alpha; N, H) = \sum_{|n^3 - N/3| \le H} e(\alpha n^4).$$

At $|p - N/3| \leq H$, using Lagrange's formula of finite increments, it is easy to show that

$$\ln p = \ln \frac{N}{3} + O\left(\frac{H}{N}\right).$$

Therefore.

$$S_1(\alpha; N, H) = \sum_{|p-N/3|\leq H} \left(\frac{\ln p}{\ln \frac{N}{3}} + O\left(\frac{H}{N\mathcal{L}}\right)\right) e(\alpha p) =$$

$$= \ln^{-1}\frac{N}{3}\left(\sum_{|n-N/3|\leq H} \Lambda(n)e(\alpha n) - \right.$$

$$\sum_{\substack{|p^k-N/3|\leq H \\ k\geq 2}} \Lambda(p^k)e(\alpha p^k)\right) + O\left(\frac{H}{N\mathcal{L}}\right) =$$

$$= \ln^{-1}\frac{N}{3}S\left(\alpha; \frac{N}{3} + H, 2H\right) + O\left(\frac{H^2}{N}\right). \tag{3.1}$$

Using the ratio

$$\left(\frac{N}{3} \pm H\right)^{\frac{1}{4}} = N_1 \pm H_1 + O(H^2 N^{-7/4}), \qquad N_1 = \sqrt[4]{\frac{N}{3}}, \qquad H_1 =$$

$$\frac{H}{4\sqrt[4]{(N/3)^3}},$$

we find

$$T_1(\alpha; N, H) = T(\alpha; N_1 + H_1, 2H_1) + O(H^2 N^{-7/4}). \tag{3.2}$$

According to Dirichlet's theorem on approximation of real numbers by rational numbers, every α from the interval $[-\text{æ}, 1 - \text{æ}]$ represent in the form

$$\alpha = \frac{a}{q} + \lambda, \qquad (a, q) = 1, \qquad 1 \leq q \leq \tau, \qquad |\lambda| \leq$$

$$\frac{1}{q\tau}. \tag{3.3}$$

In this performance. $0 \leq a \leq q - 1$ and $a = 0$ only at $q = 1$. By $\mathfrak{M}$ denote those α for which $q \leq \mathcal{L}^{40}$ in the representation (???), by $\mathfrak{m}$ - denote the remaining ones α. The set $\mathfrak{M}$ consists of non-intersecting segments. Let us divide the set $\mathfrak{M}$ into sets $\mathfrak{M}_1$ и $\mathfrak{M}_2$:

$$\mathfrak{M}_1 = \left\{\alpha: \quad \alpha \in \mathfrak{M}, \ \left|\alpha - \frac{a}{q}\right| \leq \frac{\mathcal{L}^2}{H}\right\},$$

$$\mathfrak{M}_2 = \left\{\alpha: \quad \alpha \in \mathfrak{M}, \frac{\mathcal{L}^2}{H} < \left|\alpha - \frac{a}{q}\right| \leq \frac{1}{q\tau}\right\}.$$

Let us denote by $I(\mathfrak{M}_1)$, $I(\mathfrak{M}_2)$ и $I(\mathfrak{m})$ respectively integrals over the sets $\mathfrak{M}_1$, $\mathfrak{M}_2$ и $\mathfrak{m}$. Let us have

$$I(N, H) = I(\mathfrak{M}_1) + I(\mathfrak{M}_2) + I(\mathfrak{m}).$$

In the latter formula, the first term, i.e. $I(\mathfrak{M}_1)$ delivers the main term of the asymptotic formula for $I(N, H)$, a $I(\mathfrak{M}_2)$ и $I(\mathfrak{m})$ are included in its residual term.

Calculating the integral $I(\mathfrak{M}_1)$. By definition of the integral $I(\mathfrak{M}_1)$ we have:

$$I(\mathfrak{M}_1) = \sum_{q \leq \mathcal{L}^{40}} \sum_{\substack{a=0 \\ (a,q)=1}}^{q-1} I(a, q), \tag{3.4}$$

$$I(a, q) = \int_{|\lambda| \leq \mathcal{L}^2/H} F(\alpha; N, H)e(-\alpha N)d\lambda, \tag{3.5}$$

$$F(\alpha; N, H) = S_1^2(\alpha; N, H)T_1(\alpha; N, H), \qquad \alpha = \frac{a}{q} + \lambda.$$

To the sum $S(\alpha; N/3 + H, 2H)$ let us apply lemma 2, assuming, $x = N/3 + H$, $y = 2H$, $h = HN^{-\frac{3}{4}} = N^{\frac{1}{6}}\mathcal{L}^{\frac{40}{3}}$, $b = 224$. Let us check the fulfillment of the conditions:

$$h = \frac{H}{\sqrt[4]{N^3}} = N^{\frac{1}{6}}\mathcal{L}^{\frac{40}{3}} < N^{\frac{3}{16}}\exp(-\mathcal{L}^{0,76});$$

$$h(N/3 + H)^{\frac{5}{8}}\exp\mathcal{L}^{0,76} \leq N^{\frac{1}{8}}\mathcal{L}^{\frac{40}{3}} \cdot N^{\frac{5}{8}}\exp\mathcal{L}^{0,76} \leq$$

$$\leq N^{\frac{3}{4}}\mathcal{L}^{\frac{40}{3}}\exp\mathcal{L}^{0,76} \leq N^{\frac{11}{12}}\mathcal{L}^{\frac{40}{3}} = H.$$

$$\tau = \frac{12H}{\sqrt[4]{N}} > \frac{12H}{\sqrt[4]{N}} \cdot \frac{N}{N+3H} = \frac{4H^2}{N/3+H} \cdot \frac{\sqrt[4]{N^3}}{H} = \frac{(2H)^2}{(N/3+H)h}.$$

According to this lemma, for $\alpha \in \mathfrak{M}_1$ occurs

$$S\left(\alpha; \frac{N}{3} + H, 2H\right) = \frac{\mu(q)}{\varphi(q)} \frac{\sin 2\pi\lambda H}{\pi\lambda} e\left(\frac{\lambda N}{3}\right) + R_1,$$

$$R_1 \ll H\exp(-c\ln^4\mathcal{L}).$$

Hence, taking into account the formula (???), we find

$$S_1(\alpha; N, H) = \frac{1}{\ln(N/3)} \frac{\mu(q)}{\varphi(q)} \frac{\sin 2\pi\lambda H}{\pi\lambda} e\left(\frac{\lambda N}{3}\right) + R_1.$$

Using this relation, and keeping in mind that

$$\left|\frac{\mu(q)}{\varphi(q)} \frac{\sin 2\pi\lambda H}{\pi\lambda} e\left(\frac{\lambda N}{3}\right)\right| \le \frac{H}{\varphi(q)}, \quad |S_1(\alpha; N, H)| \le \frac{H}{\mathcal{L}},$$

let's find

$$S_1^2(\alpha; N, H) = \left(\frac{\mu(q)}{\varphi(q)} \frac{\sin 2\pi\lambda H}{\pi\lambda} \frac{1}{\ln(N/3)} e\left(\frac{\lambda N}{3}\right) + R_1\right) S_1(\alpha; N, H) =$$

$$= \frac{\mu(q)}{\varphi(q)} \frac{\sin 2\pi\lambda H}{\pi\lambda} \frac{1}{\ln(N/3)} e\left(\frac{\lambda N}{3}\right) \left(\frac{\mu(q)}{\varphi(q)} \frac{\sin 2\pi\lambda H}{\pi\lambda} \frac{1}{\ln(N/3)} e\left(\frac{\lambda N}{3}\right) + R_1\right) +$$

$$R_1 S_1(\alpha; N, H) =$$

$$= \frac{\mu^2(q)}{\varphi^2(q)} \frac{\sin^2 2\pi\lambda H}{(\pi\lambda)^2 \ln^2(N/3)} e\left(\frac{2\lambda N}{3}\right) + O\left(\frac{HR_1}{\varphi(q)\mathcal{L}}\right) + O\left(\frac{HR_1}{\mathcal{L}}\right),$$

that is

$$S_1^2\left(\frac{a}{q} + \lambda; N, H\right) - \frac{\mu^2(q)}{\varphi^2(q)} \cdot \frac{\sin^2 2\pi\lambda H}{(\pi\lambda)^2 \ln^2(N/3)} \, e\left(\frac{2\lambda N}{3}\right) \ll \frac{HR_1}{\varphi(q)\mathcal{L}}.$$

Postally multiplying the obtained inequality by inequality $T_1(\alpha; N, H) \ll HN^{-3/4}$, i.e., by a trivial estimate of the sum $T_1(\alpha; N, H)$ and bearing in mind that $F(\alpha; N, H) = S_1^2(\alpha; N, H)T_1(\alpha; N, H)$ we find

$$F(\alpha; N, H) = \frac{\mu^2(q)}{\varphi^2(q)} \frac{\sin^2 2\pi\lambda H}{(\pi\lambda)^2 \ln^2(N/3)} e\left(\frac{2\lambda N}{3}\right) T_1(\alpha; N, H) +$$

$$R_2, \qquad (3.6)$$

$$R_2 \ll \frac{H^2 R_1}{\varphi(q)N^{3/4}\mathcal{L}}.$$

From the inequality

$$24(N_1 + H_1)^2 \cdot 2H_1 = 48\left(\sqrt[4]{\frac{N}{3}} + \frac{H}{4\sqrt[4]{(N/3)^3}}\right)^2 \frac{H}{4\sqrt[4]{(N/3)^3}} =$$

$$= \frac{12\sqrt[4]{3}}{\sqrt[4]{N}} H\left(1 + \frac{3H}{4N}\right)^2 < \frac{12\sqrt[4]{3}}{\sqrt[4]{N}} H\left(1 + \frac{2\sqrt[8]{3^3}-3}{3}\right)^2 = \frac{16H}{\sqrt[4]{N}} = \tau.$$

it follows that for the sum $T(\alpha; N_1 + H_1, 2H_1)$ at $\alpha \in \mathfrak{M}_1$, the condition of Corollary 2 of Theorem 2 holds. According to this corollary and the relation

(???), we find

$$T_1(\alpha; N, H) = T(\alpha; N_1 + H_1, 2H_1) + O(H^2 N^{-7/4}) =$$

$$= \frac{2S(a,q)H_1}{q} \gamma(\lambda; N_1 + H_1, 2H_1) + O(q^{1/2+\varepsilon}) + O\left(\frac{H^2}{N^{7/4}}\right),$$

where

$$S(a,q) = \sum_{x=1}^{q} e\left(\frac{ax^4}{q}\right),$$

$$\gamma(\lambda; N_1 + H_1, 2H_1) = \int_{-0,5}^{0,5} e\left(\lambda\left(\sqrt[4]{\frac{N}{3}} + \frac{Ht}{2\sqrt[4]{(N/3)^3}}\right)^4\right) dt,$$

that is

$$T_1(\alpha; N, H) = \frac{S(a,q)H}{2q\sqrt[4]{(N/3)^3}} \gamma(\lambda; N_1 + H_1, 2H_1) + R_3, \qquad R_3 \ll$$

$$q^{1/2+\varepsilon} + \frac{H^2}{N^{7/4}}.$$

Hence and from (???), we obtain

$$F(\alpha; N, H) =$$

$$= \frac{\mu^2(q)}{\varphi^2(q)} \frac{\sin^2 2\pi\lambda H}{(\pi\lambda)^2 \ln^2(N/3)} e\left(\frac{2\lambda N}{3}\right) \left(\frac{S(a,q)H}{2q\sqrt[4]{(N/3)^3}} \gamma(\lambda; N_1 + H_1, 2H_1) + \right.$$

$$\left. R_3\right) + R_2 =$$

$$= \frac{H}{2\sqrt[4]{(N/3)^3}\ln^2(N/3)} \frac{\mu^2(q)S(a,q)}{q\varphi^2(q)} \frac{\sin^2 2\pi\lambda H}{(\pi\lambda)^2} \gamma(\lambda; N_1 + $$

$$H_1, 2H_1)e\left(\frac{2\lambda N}{3}\right) + R_4,$$

$$R_4 \ll \left|\frac{\mu(q)}{\varphi(q)} \frac{\sin 2\pi\lambda H}{\pi\lambda} e\left(\frac{\lambda N}{3}\right)\right|^2 \cdot \frac{R_3}{\mathcal{L}^2} + R_2 \ll$$

$$\ll \frac{H^2 R_3}{\varphi^2(q)\mathcal{L}^2} + \frac{H^3}{\varphi(q)N^{3/4}\mathcal{L}} \exp(-c\ln^4 \mathcal{L}) \ll$$

$$\ll \frac{H^2 q^{1/2+\varepsilon}}{\varphi^2(q)\mathcal{L}^2} + \frac{H^4}{\varphi^2(q)\mathcal{L}^2 N^{7/4}} + \frac{H^3 \exp(-c\ln^4 \mathcal{L})}{\varphi(q)N^{3/4}\mathcal{L}}.$$

Substituting the value $F(\alpha; N, H)$ in formula (3.5), we find

$$I(a, q) = \frac{H}{2\sqrt[4]{(N/3)^3}\ln^2(N/3)} \frac{\mu^2(q)S(a,q)}{q\varphi^2(q)} e\left(-\frac{aN}{q}\right) \cdot J(H) +$$

$$R_5, \qquad\qquad\qquad (3.7)$$

$$J(H) = \int_{|\lambda|\leq \mathcal{L}^2/H} \frac{\sin^2 2\pi\lambda H}{(\pi\lambda)^2} \gamma(\lambda; N_1 + H_1, 2H_1) e\left(-\frac{\lambda N}{3}\right) d\lambda,$$

$$R_5 \ll \frac{Hq^{1/2+\varepsilon}}{\varphi^2(q)} + \frac{H^3}{\varphi^2(q)N^{7/4}} + \frac{H^2\mathcal{L}\exp(-c\ln^4\mathcal{L})}{\varphi(q)N^{3/4}}.$$

Substituting the value of the integral $\gamma(\lambda; N_1 + H_1, 2H_1)$ from Corollary 2 of Theorem 2 to the right-hand side of the expression for $I_1(a,q)$ we find

$$J(H) = \int_{|\lambda|\leq \mathcal{L}^2/H} \frac{\sin^2 2\pi\lambda H}{(\pi\lambda)^2} \gamma(\lambda; N_1 + H_1, 2H_1) e\left(-\frac{\lambda N}{3}\right) d\lambda =$$

$$= \int_{|\lambda|\leq \mathcal{L}^2/H} \frac{\sin^2 2\pi\lambda H}{(\pi\lambda)^2} \int_{-0,5}^{0,5} e\left(\lambda\left(\sqrt[4]{\frac{N}{3}} + \right.\right.$$

$$\left.\left. \frac{Ht}{2\sqrt[4]{(N/3)^3}}\right)^4\right) e\left(-\frac{\lambda N}{3}\right) dt \quad d\lambda =$$

$$= \int_{|\lambda|\leq \mathcal{L}^2/H} \frac{\sin^2 2\pi\lambda H}{(\pi\lambda)^2} \int_{-0,5}^{0,5} e\left(2\lambda Ht\left(1 + \frac{9Ht}{4N} + \frac{9H^2t^2}{4N^2} + \right.\right.$$

$$\left.\left. \frac{27H^3t^3}{32N^3}\right)\right) dt \quad d\lambda =$$

$$= \frac{2H}{\pi} \int_{|u|\leq 2\pi\mathcal{L}^2} \frac{\sin^2 u}{u^2} \int_{-0,5}^{0,5} e\left(\frac{ut}{\pi}\left(1 + \frac{9Ht}{4N} + \frac{9H^2t^2}{4N^2} + \right.\right.$$

$$\left.\left. \frac{27H^3t^3}{32N^3}\right)\right) dt \quad du.$$

Using the ratio

$$e\left(\frac{ut}{\pi}\left(\frac{9Ht}{4N} + \frac{9H^2t^2}{4N^2} + \frac{27H^3t^3}{32N^3}\right)\right) =$$

$$= e\left(\frac{9Hut^2}{4\pi N}\left(1 + \frac{Ht}{N} + \frac{3H^2t^2}{8N^2}\right)\right) = 1 +$$

$$O\left(\frac{H\mathcal{L}^2}{N}\right).$$

let's find

$$J(H) =$$

$$\frac{2H}{\pi} \int_{|u|\leq 2\pi\mathcal{L}^2} \frac{\sin^2 u}{u^2} \int_{-0,5}^{0,5} e\left(\frac{ut}{\pi}\right) e\left(\frac{ut}{\pi}\left(\frac{9Ht}{4N} + \frac{9H^2 t^2}{4N^2} + \right.\right.$$

$$\left.\left.\frac{27H^3 t^3}{32N^3}\right)\right) dt \quad du =$$

$$= \frac{2H}{\pi} \int_{|u|\leq 2\pi\mathcal{L}^2} \frac{\sin^2 u}{u^2} \int_{-0,5}^{0,5} e\left(\frac{ut}{\pi}\right)\left(1 + O\left(\frac{H\mathcal{L}^2}{N}\right)\right) du \quad dt =$$

$$= \frac{2H}{\pi} \int_{|u|\leq 2\pi\mathcal{L}^2} \frac{\sin^2 u}{u^2} \int_{-0,5}^{0,5} e\left(\frac{ut}{\pi}\right) dt \quad du + O\left(\frac{H^2 \mathcal{L}^4}{N}\right) =$$

$$= \frac{2H}{\pi} \int_{|u|\leq 2\pi\mathcal{L}^2} \frac{\sin^3 u}{u^3} du + O\left(\frac{H^2 \mathcal{L}^4}{N}\right) = \frac{4H}{\pi} \int_0^{2\pi\mathcal{L}^2} \frac{\sin^3 u}{u^3} du +$$

$$O\left(\frac{H^2 \mathcal{L}^4}{N}\right).$$

Replacing the last integral by u by a close to it non-similar integral independent of $\mathcal{L}$ and using the relation

$$\int_{2\pi\mathcal{L}^2}^{\infty} \frac{\sin^3 u}{u^3} du \ll \mathcal{L}^{-6},$$

let's find

$$J(H) = \frac{4H}{\pi}\left(\int_0^{\infty} \frac{\sin^3 u}{u^3} du - \int_{2\pi\mathcal{L}^2}^{\infty} \frac{\sin^3 u}{u^3} du\right) + O\left(\frac{H^2 \mathcal{L}^4}{N}\right) =$$

$$= \frac{4H}{\pi} \int_0^{\infty} \frac{\sin^3 u}{u^3} du + O\left(\frac{H}{\mathcal{L}^6}\right).$$

Using the formula (see [44] p. 174)

$$\int_0^{\infty} \frac{\sin^n mu}{u^n} du$$

$$= \frac{\pi m^{m-1}}{2^n (n-1)1}\left[n^{n-1} - \frac{n}{1!}(n-2)^{n-1} + \frac{n(n-1)}{2!}(n-4)^{n-1}\right.$$

$$\left. + \cdots\right].$$

at $m = 1$ и $n - 3$ we find

$$\int_0^{\infty} \frac{\sin^3 u}{u^3} du = \frac{3\pi}{8}.$$

Hence.

$$J(H) = \frac{3H}{2} + O\left(\frac{H}{\mathcal{L}^6}\right).$$

Substituting this into formula (3.7.), we find

$$I(a,q) = \frac{H}{2\sqrt[4]{(N/3)^3}\ln^2(N/3)} \cdot \frac{\mu^2(q)S(a,q)}{q\varphi^2(q)} e\left(-\frac{aN}{q}\right) \cdot$$

$$\left(\frac{3H}{2} + O\left(\frac{H}{\mathcal{L}^6}\right)\right) + R_5 =$$

$$= \frac{3H}{4\sqrt[4]{(N/3)^3}\ln^2(N/3)} \cdot \frac{\mu^2(q)S(a,q)}{q\varphi^2(q)} e\left(-\frac{aN}{q}\right) + R_6(a,q), \qquad (3.8)$$

$$R_6(a,q) \ll \frac{\mu^2(q)|S(a,q)|}{q\varphi^2(q)} \cdot \frac{H}{\sqrt[4]{N^3}\mathcal{L}^2} \cdot \frac{H}{\mathcal{L}^6} + R_5.$$

If q - is quadratic, then according to Lemma 2.1, for a complete rational sum of order four $S(a,q)$ is valid for $S(a,q) \ll \sqrt{q}$. Using this estimate and the explicit expression R_5 we find

$$R_6(a,q) \ll \frac{1}{\sqrt{q}\varphi^2(q)} \frac{H^2}{\sqrt[4]{N^3}\mathcal{L}^8} + \frac{Hq^{1/2+\varepsilon}}{\varphi^2(q)} + \frac{H^3}{\varphi^2(q)N^{7/4}} +$$

$$\frac{H^2\mathcal{L}\exp(-c\ln^4\mathcal{L})}{\varphi(q)N^{3/4}}.$$

Substituting the right part of the equality (??) into (??), we obtain

$$I(\mathfrak{M}_1) =$$

$$\frac{3H\ln^{-2}(N/3)}{4\sqrt[4]{(N/3)^3}} \sum_{q\leq\mathcal{L}^{40}} \frac{\mu^2(q)}{q\varphi^2(q)} \sum_{\substack{a=0\\(a,q)=1}}^{q-1} S(a,q)e\left(-\frac{aN}{q}\right) +$$

$$R(\mathfrak{M}_1), \qquad (3.9)$$

$$R(\mathfrak{M}_1) \ll \sum_{q\leq\mathcal{L}^{40}} \sum_{\substack{a=0\\(a,q)=1}}^{q-1} R_6(a,q) \ll$$

$$\frac{H^2}{\sqrt[4]{N^3}\mathcal{L}^8} \sum_{q\leq\mathcal{L}^{40}} \sum_{\substack{a=0\\(a,q)=1}}^{q-1} \frac{1}{\sqrt{q}\varphi^2(q)} +$$

$$+H \sum_{q\leq\mathcal{L}^{40}} \sum_{\substack{a=0\\(a,q)=1}}^{q-1} \frac{q^{1/2+\varepsilon}}{\varphi^2(q)} +$$

$$\frac{H^3}{N^{7/4}} \sum_{q\leq\mathcal{L}^{40}} \sum_{\substack{a=0\\(a,q)=1}}^{q-1} \frac{1}{\varphi^2(q)} +$$

$$+ \frac{H^2 \mathcal{L}}{N^{3/4}} \exp(-c\ln^4 \mathcal{L}) \sum_{q \leq \mathcal{L}^{40}} \sum_{\substack{a=0 \\ (a,q)=1}}^{q-1} \frac{1}{\varphi(q)} \ll$$

$$\ll \frac{H^2}{\sqrt[4]{N^3}\mathcal{L}^8} \sum_{q \leq \mathcal{L}^{40}} \frac{1}{q^{1/2}\varphi(q)} + H \sum_{q \leq \mathcal{L}^{40}} \frac{q^{1/2+\varepsilon}}{\varphi(q)} +$$

$$+ \frac{H^3}{N^{7/4}} \sum_{q \leq \mathcal{L}^{40}} \frac{1}{\varphi(q)} + \frac{H^2 \mathcal{L}^{41}}{N^{3/4}} \exp(-c\ln^4 \mathcal{L}) \ll$$

$$\ll \frac{H^2}{\sqrt[4]{N^3}\mathcal{L}^8} + H\mathcal{L}^{21} + \frac{H^3}{N^{7/4}}\mathcal{L}^2 + \frac{H^2}{\sqrt[4]{N^3}\mathcal{L}^8} \ll \frac{H^2}{\sqrt[4]{N^3}\mathcal{L}^8}.$$

Thus,

$$I(\mathfrak{M}_1) =$$

$$\frac{3H}{4\sqrt[4]{(N/3)^3}\ln^2(N/3)} \sum_{q \leq \mathcal{L}^{40}} \frac{\mu^2(q)}{q\varphi^2(q)} \sum_{\substack{a=0 \\ (a,q)=1}}^{q-1} S(a,q)e\left(-\frac{aN}{q}\right) +$$

$$O\left(\frac{H^2}{\sqrt[4]{N^3}\mathcal{L}^8}\right). \quad (3.10)$$

The sum over q in (???) let us replace it by a close to it infinite series independent of $\mathcal{L}^{40}$ that is

$$\sum_{q \leq \mathcal{L}^{40}} \frac{\mu^2(q)}{q\varphi^2(q)} \sum_{\substack{a=1 \\ (a,q)=1}}^{q} S(a,q)e(-aN/q) = \mathfrak{S}(N) -$$

$$R(N), \qquad (3.11)$$

$$\mathfrak{S}(N) = \sum_{q=1}^{\infty} \frac{\mu^2(q)}{q\varphi^2(q)} \sum_{\substack{a=1 \\ (a,q)=1}}^{q} S(a,q)e\left(-\frac{aN}{q}\right),$$

$$R(N) = \sum_{q > \mathcal{L}^{40}} \frac{\mu^2(q)}{q\varphi^2(q)} \sum_{\substack{a=1 \\ (a,q)=1}}^{q} S(a,q)e\left(-\frac{aN}{q}\right).$$

Now let us represent the special series $\mathfrak{S}(N)$ as an infinite product of all prime numbers, for this purpose let us write it in the form

$$\mathfrak{S}(N) = \sum_{q=1}^{\infty} \frac{\mu^2(q)}{q\varphi^2(q)} \Phi(q), \qquad \Phi(q) =$$

$$\sum_{\substack{a=1 \\ (a,q)=1}}^{q} S(a,q)e\left(-\frac{aN}{q}\right).$$

Let us first show that the sum $\Phi(q)$ is a multiplicative function. Let $q = q_1 q_2$, $(q_1, q_2) = 1$, then representing the summation variable in the last sum

42

a in the form

$$a = a_1 q_2 + a_2 q_1, \qquad (a_1, q_1) = 1, \qquad 1 \leq a_1 \leq$$

$q_1, \qquad (a_2, q_2) = 1, \qquad 1 \leq a_2 \leq q_2,$

let's find

$$\Phi(q_1 q_2) = \sum_{\substack{a_1=1 \\ (a_1,q_1)=1}}^{q_1} \sum_{\substack{a_2=1 \\ (a_2,q)=1}}^{q_2} S(a_1 q_2 +$$

$$a_2 q_1, q_1 q_2) e\left(-\frac{(a_1 q_2 + a_2 q_1)N}{q_1 q_2}\right). \tag{3.12}$$

Applying to the sum $S(a_1 q_2 + a_2 q_1, q_1 q_2)$ lemma 2, we have

$$S(a_1 q_2 + a_2 q_1, q_1 q_2) = S(a_1, q_1)S(a_2, q_2).$$

Substituting this equality into the right-hand side (???), we obtain

$$\Phi(q_1 q_2) =$$

$$\sum_{\substack{a_1=1 \\ (a_1,q_1)=1}}^{q_1} S(a_1, q_1) e\left(-\frac{a_1 N}{q_1}\right) \sum_{\substack{a_2=1 \\ (a_2,q)=1}}^{q_2} S(a_2, q_2) e\left(-\frac{a_2 N}{q_2}\right) =$$

$$\Phi(q_1)\Phi(q_2).$$

Using the absolute convergence $\mathfrak{S}(N)$ and multiplicativity $\Phi(q)$ we find

$$\mathfrak{S}(N) = \prod_p \left(1 + \frac{\Phi(p)}{p(p-1)^2}\right),$$

$$\Phi(p) = \sum_{a=1}^{p-1} S(a,p) e\left(-\frac{aN}{p}\right) = \sum_{a=1}^{p-1} \sum_{x=1}^{p} e\left(\frac{a(x^4-N)}{p}\right) =$$

$$= \sum_{x=1}^{p} \sum_{a=1}^{p-1} e\left(\frac{a(x^4-N)}{p}\right) = \sum_{x=1}^{p} \left(\sum_{a=1}^{p} e\left(\frac{a(x^4-N)}{p}\right) -$$

$$1\right) =$$

$$= \sum_{\substack{x=1 \\ x^3 \equiv N(mod p)}}^{p} p - p = p(\rho(N,p) - 1)),$$

where $\rho(N,p)$ - is the number of comparison solutions $x^4 \equiv N(mod p)$.

Hence,

$$\mathfrak{S}(N) = \prod_p \left(1 + \frac{\rho(N,p)-1}{(p-1)^2}\right).$$

Now, let's evaluate $R(N)$. We have

$$R(N) = \sum_{q > \mathcal{L}^{40}} \frac{\mu^2(q)}{q\varphi^2(q)} \Phi(q), \qquad \Phi(q) =$$

$$\sum_{\substack{a=1 \\ (a,q)=1}}^{q} S(a,q) e\left(-\frac{aN}{q}\right).$$

Since $\Phi(q)$ - multiplicative function and q - is a squareless number, then

$$\Phi(q) = \prod_{p\backslash q} \Phi(p), \qquad \Phi(p) =$$

$$\sum_{a=1}^{p-1} S(a,p) e\left(-\frac{aN}{p}\right).$$

Hence and from the estimation $|S(a,p)| \ll p^{1/2}$ (lemma 2), successively we find

$$|\Phi(q)| = \prod_{p\backslash q} |\Phi(p)| \le \prod_{p\backslash q} \sqrt{p}(p-1) < \prod_{p\backslash q} p\sqrt{p} = $$

$$q\sqrt{q}.$$

Consequently

$$|R(N)| \le \sum_{q > \mathcal{L}^{40}} \frac{\mu^2(q)}{q\varphi^2(q)} |\Phi(q)| \le \sum_{q > \mathcal{L}^{40}} \frac{\mu^2(q)}{q\varphi^2(q)} \cdot q\sqrt{q} =$$

$$= \sum_{q > \mathcal{L}^{40}} \frac{\mu^2(q)}{\sqrt{q}\varphi(q)} \ll \sum_{q > \mathcal{L}^{40}} \frac{\ln\ln q}{q^{3/2}} \ll \int_{\mathcal{L}^{40}}^{\infty} \frac{\ln\ln u}{u^{3/2}} du \ll \mathcal{L}^{-8}.$$

Thus the relation (???) takes the form

$$\sum_{q \le Q} \frac{\mu^2(q)}{q\varphi^2(q)} \sum_{\substack{a=1 \\ (a,q)=1}}^{q} S(a,q) e\left(-\frac{aN}{q}\right) = \mathfrak{S}(N) + O(\mathcal{L}^{-8}),$$

$$\mathfrak{S}(N) = \prod_{p} \left(1 + \frac{\rho(N,p)-1}{(p-1)^2}\right),$$

where $\rho(N,p)$ - is the number of comparison solutions $x^4 \equiv N(mod\,p)$.

Substituting the right part of this equality into (??), we obtain

$$I(\mathfrak{M}_1) = \frac{3H\mathfrak{S}(N)}{4\sqrt[4]{(N/3)^3}\ln^2(N/3)} + O\left(\frac{H^2}{\sqrt[4]{N^3}\mathcal{L}^8}\right).$$

Hence, taking into account the formula $\ln^2(N/3) = \mathcal{L}^{-2} + O(\mathcal{L}^{-3})$ we find

$$I(\mathfrak{M}_1) = \frac{\sqrt[4]{3}\mathfrak{S}(N)}{4\sqrt[4]{N^3}\mathcal{L}^2} H + O\left(\frac{H^2}{\sqrt[4]{N^3}\mathcal{L}^2}\right). \tag{3.13}$$

Evaluating the integral $I(\mathfrak{M}_2)$. We have

$$I(\mathfrak{M}_1) = \int_{\mathfrak{m}} S_1^2(\alpha; N, H) T_1(\alpha; N, H) e(-\alpha N) d\alpha.$$

Turning to estimates, we find

$$I(\mathfrak{M}_1) \ll \max_{\alpha \in \mathfrak{m}} |T_1(\alpha, N, H)| \int_0^1 |S_1(\alpha; N, H)|^2 d\alpha =$$

$$= \max_{\alpha \in \mathfrak{M}_1} |T_1(\alpha, N, H)| \int_0^1 S_1(\alpha; N, H) \overline{S_1(\alpha; N, H)} d\alpha =$$

$$= \max_{\alpha \in \mathfrak{M}_1} |T_1(\alpha, N, H)| \sum_{|p-N/3| \leq H} 1 =$$

$$= \max_{\alpha \in \mathfrak{M}_1} |T_1(\alpha, N, H)| \left(\pi\left(\frac{N}{3} + H\right) - \pi\left(\frac{N}{3} - H\right) \right).$$

Applying to the right part of the obtained formula taking into account

the ratio $y \gg x^{\frac{11}{12}} \mathcal{L}^{\frac{40}{3}} \geq x^{\frac{7}{12}+\varepsilon}$ lemma 2, we find

$$I(\mathfrak{M}_1) \ll \frac{H}{\mathcal{L}} \cdot \max_{\alpha \in \mathfrak{M}_2} |T_1(\alpha, N, H)|. \qquad (3.14)$$

If $\alpha \in \mathfrak{M}_2$ then

$$\alpha = \frac{a}{q} + \lambda, \qquad (a, q) = 1, \qquad \frac{\mathcal{L}^2}{H} < |\lambda| \leq \frac{1}{q\tau}, \qquad 1 \leq q \leq \mathcal{L}^{40}.$$

Let's estimate $T_1(\alpha, N, H)$ for α from the set $\mathfrak{M}_2$. According to the relation (???), we have

$$T_1(\alpha; N, H) = T(\alpha; N_1 + H_1, 2H_1) + O(H^2 N^{-7/4}), \quad (3.15)$$

$$N_1 = \sqrt[4]{\frac{N}{3}}, \qquad H_1 = \frac{H}{4\sqrt[4]{(N/3)^3}}.$$

Let's estimate $T(\alpha, N_1 + H_1, 2H_1)$ for α from the set $\mathfrak{M}_2$. Let us consider two possible cases:

1. $\dfrac{\mathcal{L}^2}{H} < |\lambda| \leq \dfrac{1}{8q(N_1+H_1)^3}$;

2. $\dfrac{1}{8q(N_1+H_1)^3} < |\lambda| \leq \dfrac{1}{q\tau}$.

Case 1. Earlier, when calculating the integral $I(\mathfrak{M}_1)$ we showed that

$$\tau = \frac{16H}{\sqrt[4]{N}} > 24(N_1 + H_1)^2 \cdot 2H_1.$$

Hence, in this case for the sum $T(\alpha; N_1 + H_1, 2H_1)$ и $\alpha \in \mathfrak{M}_2$, the condition of Corollary 2 of Theorem 2 is satisfied. According to the corollary, we have

$$T(\alpha, N_1 + H_1, 2H_1) = \frac{2H_1}{q} S(a, q)\gamma(\lambda; N_1 + H_1, 2H_1) + O(q^{1/2+\varepsilon}).$$

$$(3.16)$$

Let us apply to the evaluation of the integral

$$\gamma(\lambda; N_1 + H_1, 2H_1) = \int_{-0.5}^{0.5} e(\lambda(N_1 + 2H_1 u)^4)du$$

lemma 1, assuming $f(u) = \lambda(N_1 + 2H_1 u)^4$. The second order derivative $f''(u) = 48\lambda H_1^2(N_1 + 2H_1 u)^2$ does not change sign, hence, the first order derivative $f'(u) = 8\lambda H_1(N_1 + 2H_1 u)^3$ is a monotone function and satisfies the inequality

$$|f'(u)| \geq 8|\lambda|H_1(N_1 - H_1)^3 = 8|\lambda|H_1(N_1^3 - 3N_1^2 H_1 +$$

$$3N_1 H_1^2 - H_1^3) >$$

$$> 8|\lambda|(N_1^3 H_1 - 3N_1^2 H_1^2 - H_1^4) =$$

$$= 6|\lambda| \left(\sqrt[4]{(N/3)^3} \cdot \frac{H}{4\sqrt[4]{(N/3)^3}} - 3\sqrt[4]{(N/3)^2} \cdot \frac{H^2}{16\sqrt[4]{(N/3)^6}} - \right.$$

$$\left. \frac{H^4}{4\sqrt[4]{(N/3)^{12}}} \right) =$$

$$= 6|\lambda| \left(\frac{H}{4} - \frac{9H^2}{16N} - \frac{27H^4}{4N^3} \right) = \frac{3|\lambda|H}{2} \left(1 - \frac{9H}{4N} - \frac{27H^3}{N^3} \right) > |\lambda|H \geq$$

$$\mathcal{L}^2.$$

Hence, according to lemma 1, at $m = \mathcal{L}^2$ и $M = 1$ we find

$$|\gamma(\lambda; N_1 + H_1, 2H_1)| \leq \mathcal{L}^{-2}.$$

Therefore, passing in (???) to estimates and using lemma 1 on the estimation of the full rational trigonometric sum $S(a, q)$ we find

$$|T(\alpha, N_1 + H_1, 2H_1)| \ll \frac{H_1|S(a,q)||\gamma(\lambda;N_1+H_1,2H_1)|}{q} + q^{\frac{1}{2}+\varepsilon} \ll$$

$$\ll \frac{H}{q^{1/4}\sqrt[4]{N^3}\mathcal{L}^2} + \mathcal{L}^{20+40\varepsilon} \ll \frac{H}{\sqrt[4]{N^3}\mathcal{L}^2} + \mathcal{L}^{21} \ll \frac{H}{\sqrt[4]{N^3}\mathcal{L}^2}. \quad (3.17)$$

Case 2. In this case, i.e. at

$$\frac{1}{8q(N_1+H_1)^3} < |\lambda| \le \frac{1}{q\tau},$$

according to Corollary 2 of Theorem 2 when $x = N_1 + H_1$, $y = 2H_1$ we have

$$T(\alpha, N_1 + H_1, 2H_1) \ll q^{\frac{3}{4}}\ln q + \min_{2\le k\le 4}\left(H_1 q^{-\frac{1}{4}}, (N_1 + H_1)^{1-\frac{1}{k}}q^{\frac{1}{k}-\frac{1}{4}}\right) \le$$

$$\le q^{\frac{3}{4}}\ln q + (N_1 + H_1)^{\frac{1}{2}}q^{\frac{1}{2}} \ll$$

$$\ll \mathcal{L}^{30}\ln\mathcal{L} + \left(\sqrt[4]{\frac{N}{3}} + \frac{H}{4\sqrt[4]{(N/3)^3}}\right)^{\frac{1}{2}}\mathcal{L}^{20} \ll$$

$$\ll \mathcal{L}^{30}\ln\mathcal{L} + \left(N^{\frac{1}{4}} + N^{\frac{1}{6}}\mathcal{L}^{\frac{40}{3}}\right)^{\frac{1}{2}}\mathcal{L}^{20} \ll N^{\frac{1}{8}}\mathcal{L}^{20} =$$

$$= \frac{H}{\sqrt[4]{N^3}\mathcal{L}^2} \cdot \frac{N^{\frac{7}{8}}\mathcal{L}^{22}}{H} \le \frac{H}{\sqrt[4]{N^3}\mathcal{L}^2}.$$

Hence and from the estimation (3.17), taking into account the relation (???) for all $\alpha \in \mathfrak{M}_2$ we obtain

$$|T_1(\alpha; N, H)| \ll |T(\alpha; N_1 + H_1, 2H_1)| + H^2 N^{-7/4} \ll$$

$$\ll \frac{H}{\sqrt[4]{N^3}\mathcal{L}^2}\left(1 + \frac{H\mathcal{L}^2}{N}\right) \ll \frac{H}{\sqrt[4]{N^3}\mathcal{L}^2}.$$

Substituting the obtained estimate for $|T_1(\alpha; N, H)$, $\alpha \in \mathfrak{M}_2$, in (???), we obtain

$$I(\mathfrak{M}_2) \ll \frac{H}{\mathcal{L}} \cdot \max_{\alpha \in \mathfrak{M}_2}|T_1(\alpha, N, H)| \ll \frac{H^2}{\sqrt[4]{N^4}\mathcal{L}^3}.$$

Evaluating the integral $I(\mathfrak{m})$. We have

$$I(\mathfrak{m}) = \int_{\mathfrak{m}} S_1^2(\alpha; N, H)T_1(\alpha; N, H)e(-\alpha N)d\alpha.$$

Turning to estimates, we find

$$I(m) \ll \max_{\alpha \in m}|T_1(\alpha, N, H)| \int_0^1 |S_1(\alpha; N, H)|^2 d\alpha =$$

$$= \max_{\alpha \in m}|T_1(\alpha, N, H)| \int_0^1 S_1(\alpha; N, H)\overline{S_1(\alpha; N, H)}d\alpha =$$

$$= \max_{\alpha \in m}|T_1(\alpha, N, H)| \sum_{|p-N/3|\leq H} 1 =$$

$$= \max_{\alpha \in m}|T_1(\alpha, N, H)| \left(\pi\left(\frac{N}{3} + H\right) - \pi\left(\frac{N}{3} - H\right) \right).$$

Applying to the right part of the obtained formula taking into account the ratio $y \gg x^{\frac{11}{12}}\mathcal{L}^{\frac{40}{3}} \geq x^{\frac{7}{12}+\varepsilon}$ lemma 2, we find

$$I(m) \ll \frac{H}{\mathcal{L}} \cdot \max_{\alpha \in \mathfrak{M}_2} |T_1(\alpha, N, H)|. \tag{3.18}$$

If $\alpha \in m$ then

$$\alpha = \frac{a}{q} + \lambda, \qquad (a, q) = 1, \qquad |\lambda| \leq \frac{1}{q\tau}, \qquad \mathcal{L}^{40} \leq q \leq \tau.$$

Let's estimate $T_1(\alpha, N, H)$ for α from the set m. Recall that the sums $T_1(\alpha; N, H)$ и $T(\alpha; N_1 + H_1, 2H_1)$ are related by the formula

$$T_1(\alpha; N, H) = T(\alpha; N_1 + H_1, 2H_1) + O(H^2 N^{-\frac{7}{4}}), \tag{3.19}$$

$$N_1 = \sqrt[4]{\frac{N}{3}}, \qquad\qquad H_1 = \frac{H}{4\sqrt[4]{(N/3)^3}}.$$

Using for the sum $T(\alpha; N_1 + H_1, 2H_1)$ Theorem 3 and the condition

$$\mathcal{L}^{40} < q \leq \tau = 16HN^{-\frac{1}{4}},$$

we find

$$T(\alpha; N_1, +H_1, 2H_1) \ll H_1 \left(q^{-\frac{1}{16}} + H_1^{-\frac{1}{16}}\ln^{\frac{1}{16}}q + \right.$$

$$\left. H_1^{-\frac{1}{4}}q^{\frac{1}{16}}\ln^{\frac{1}{16}}q \right) (\ln H_1)^{\frac{7}{16}} \ll$$

$$\ll \frac{H}{\sqrt[4]{N^3}} \left(\mathcal{L}^{-\frac{5}{2}} + \left(\frac{H}{\sqrt[4]{N^3}}\right)^{-\frac{1}{16}} \mathcal{L}^{\frac{1}{16}} + \left(\frac{H}{\sqrt[4]{N^3}}\right)^{-\frac{1}{4}} \left(\frac{H}{\sqrt[4]{N}}\right)^{\frac{1}{16}} \mathcal{L}^{\frac{1}{16}} \right) \mathcal{L}^{\frac{7}{16}} =$$

$$= \frac{H}{\sqrt[4]{N^3}} \left(\mathcal{L}^{-\frac{33}{16}} + \left(\frac{H}{\sqrt[4]{N^3}} \right)^{-\frac{1}{16}} \mathcal{L}^{\frac{1}{2}} + \left(\frac{N^{\frac{11}{12}}}{H} \right)^{\frac{3}{16}} \mathcal{L}^{\frac{1}{2}} \right) =$$

$$= \frac{H}{\sqrt[4]{N^3}} \left(\mathcal{L}^{-\frac{33}{16}} + \left(\frac{N^{\frac{11}{12}}\mathcal{L}^{\frac{40}{3}}}{\sqrt[4]{N^3}} \right)^{-\frac{1}{16}} \mathcal{L}^{\frac{1}{2}} + \left(\frac{N^{\frac{11}{12}}}{N^{\frac{11}{12}}\mathcal{L}^{\frac{40}{3}}} \right)^{\frac{3}{16}} \mathcal{L}^{\frac{1}{2}} \right) =$$

$$= \frac{H}{\sqrt[4]{N^3}} \left(\mathcal{L}^{-\frac{33}{16}} + N^{-\frac{1}{96}}\mathcal{L}^{-\frac{1}{3}} + \mathcal{L}^{-2} \right) \ll \frac{H}{\sqrt[4]{N^3}\mathcal{L}^2}.$$

Hence and from (???), we obtain

$$|T_1(\alpha; N, H)| \ll |T(\alpha; N_1 + H_1, 2H_1)| + \frac{H^2}{N^{\frac{7}{4}}} \ll \frac{H}{\sqrt[4]{N^3}\mathcal{L}^2} + \frac{H^2}{N^{\frac{7}{4}}} =$$

$$= \frac{H}{\sqrt[4]{N^3}\mathcal{L}^2} \left(1 + \frac{H}{N} \right) \ll \frac{H}{\sqrt[4]{N^3}\mathcal{L}^2}.$$

Substituting the obtained estimate for $|T_1(\alpha; N, H)|$, $\alpha \in \mathfrak{m}$, in (???), we obtain

$$I(\mathfrak{m}) \ll \frac{H}{\mathcal{L}} \cdot \max_{\alpha \in \mathfrak{M}_2} |T_1(\alpha, N, H)| \ll \frac{H^2}{\sqrt[4]{N^4}\mathcal{L}^3}.$$

The theorem is proved.

<u>Conclusion</u>

The main results of the thesis are novel, they are substantiated by detailed evidence and are as follows:

- the behavior of short trigonometric sums of G. Weyl in large arcs is studied;

- a nontrivial estimate of short Weyl trigonometric sums of fourth order in small arcs is found;

- we prove an asymptotic formula for the number of representations of a sufficiently large natural number as a sum of three almost equal summands, two of which are prime numbers and the third is the fourth degree of the natural number.

The results obtained in the thesis are of theoretical nature, its results and the methodology of their derivation can be used by specialists in the field of analytic number theory.

Literature

[1] Vinogradov, I.M. Selected works / I.M. Vinogradov //Moscow: Publishing House of the USSR Academy of Sciences. 1952.

[2] Haselgrove C.B.. Some theorems in the analitic theory of number / C.B.Haselgrove // J.London Math.Soc.,26 (1951). - P. 273 - 277.

[3] Statuļavičius V.A. On the representation of odd numbers by the sum of three almost equal prime numbers / V.A.Statuļavičius // Vilnius, University Papers. ser. mat., phys. and chem. n.,3 (1955). - C. 5 - 23.

[4] Jia Chaohua Three primes theorem in a short interval (II) / Chaohua Jia // International symposium in memory of Hua Loo Keng, Science Press and Springer-Verlag, Berlin, 1991. - P. 103 - 115.

[5] Jia Chaohua Three primes theorem in a short interval (V) / Chaohua Jia // Acta Math. Sin., New Series, 2(1991). - P. 135 - 170.

[6] Jia Chaohua Three primes theorem in a short interval (VII) / Chaohua Jia // Acta Math. Sin., New Series, 10(1994). - P. 369 - 387.

[7] Jia Chaohua Three primes theorem in a short interval (VII) / Chaohua Jia // Acta Math. Sinica 4(1994). - P. 464 - 473, Chinese.

[8] Pan Cheng-dong On estimations of trigonometric sums over primes in short intervals (III) / Pan Cheng-dong, Pan Cheng-biao // Chinese Ann. of Math. of Math., 2(1990). - P. 138 - 147.

[9] Zhan Tao On the Representation of large odd integer as a sum of three almost equal primes / T.Zhan // Acta Math Sinica, new ser., 7 (1991), No 3. - P. 135 - 170.

[10] Jia Chao-hua Three primes theorem in a short interval (VII) / Jia Chao-hua // Acta Mathematica Sinica, New Series 1994. V. 10, 4. - P. 369 - 387.

[11] J Y Liu. On sums of five almost equal prime squares / J Y Liu, T. Zhan // Acta Arithmetica, 1996, 77:. - P. 369 - 383

[12] J Y Liu. On sums of five almost equal prime squares (II) / J Y Liu, T. Zhan // Sci China, 1998, 41:. - P. 710 - 722

[13] J Y Liu. Estimation of exponential sums over primes in short intervals I / J Y Liu, T. Zhan // Mh Math, 1999, 127:. - P. 27 - 41

[14] J Y Liu. Hua's Theorem on Prime Squares in Short Intervals / J Y Liu, T. Zhan // Acta Mathematica Sinica, English Series Oct., 2000, Vol. 16, No. 4. - P. - P. 669 - 690.

[15] Hua L.K.. Some results in the additive prime number theory / L.K.Hua // Quart J Math (Oxford). 1938. V. 9, No 1 . - P. 68 -- 80.

[16] Kumchev A V. On Weyl sums over primes in short intervals / A.V.Kumchev // "Arithmetic in Shangrila"-Proceedings of the 6th China-Japan Seminar on Number Theory. Series on Number Theory and Its Applications. 2012. V. 9. Singapore: World Scientific. - P. 116-131.

[17] *Yao Y.* Sums of nine almost equal prime cubes // Frontiers of Mathematics in China. October 2014. V. 9, Is. 5. - P. 1131 - 1140. DOI:10.1007/s11464-014-0384-4.

[Rakhmonov, Z.H. Ternary problem of Esterman with almost equal summands / Z.H. Rakhmonov // Mathematical Notes. 2003. - V. 74. vol. 4. - C. 564 - 572.

[19] Rakhmonov Z.H. Short quadratic trigonometric Weyl sums / Z.H. Rakhmonov , J.A. Shokamolova // Proceedings of the Academy of Sciences of the Republic of Tajikistan. Department of physico-mathematical, chemical, geological and technical sciences. 2009.   2(135). - C. 7 - 18.

[20] Rakhmonov, Z.H. Cubic problem of Esterman with almost equal summands / Z.H. Rakhmonov // Mat.notes. 2014.   - Vol. 95. vol. 3. - P. 445 - 456.

[21] Rakhmonov, Z.H. On estimates of short cubic sums of G. Weyl / Z.H. Rakhmonov, K.I. Mirzoabdugafurov // Reports of the Academy of Sciences of the Republic of Tajikistan. 2008.   - T. 51. 1. C. 5 - 15.

[22] Rakhmonov, Z.H. Estimation of short trigonometric sums of G. Weyl of the fourth degree / Z.H. Rakhmonov, A.Z. Azamov, K.I. Mirzoabdugafurov // Reports of the Academy of Sciences of the Republic of Tajikistan. 2010.   - T. 53.   10. - C. 737 - 744.

[23] Shokamolova J.A. Asymptotic formula in the Esterman problem with almost equal summands / J.A. Shokamolova // Reports of the

Academy of Sciences of the Republic of Tajikistan, 2010. - T. 53, 5. - C. 325-332.

[24] Rakhmonov, Z.H. About one ternary problem with almost equal summands / Z.H. Rakhmonov, D.M. Fozilova // Reports of the Academy of Sciences of the Republic of Tajikistan. 2012. - T. 55.   6. - C. 433 - 440.

[25] Rakhmonov, Z.H. The Varing problem for cubes with almost equal summands / Z.H. Rakhmonov, K.I. Mirzoabdugafurov // Reports of the Academy of Sciences of the Republic of Tajikistan, 2008. - T. 51.   2. - C. 83 - 86.

[26] Rakhmonov, Z.H. Asymptotic formula in the Varing problem for fourth powers with almost equal summands / Z.H. Rakhmonov, A.Z. Azamov // Reports of the Academy of Sciences of the Republic of Tajikistan. 2011. - T. 54.   3. - C. 34 - 42.

[27] Rakhmonov, Z.H. The Varing problem for fifth powers with almost equal summands / Z.H. Rakhmonov, N.N. Nazrubloev // Reports of the Academy of Sciences of the Republic of Tajikistan. 2014. - T. 57.   11 - 12. - C. 823 - 830.

[28] Fatkina, S.Yu. On one generalization of the ternary Goldbach problem for almost equal summands / S.Yu.Fatkina // Bulletin of Moscow University. series 1, mathematics. mechanics. 2001.   2

[Rakhmonov, P.Z. Short sums with non-integer degree of a natural number / P.Z.Rakhmonov // Mathematical Notes. 2014. - T. 95.   5. - C. 763 - 774.

[Rakhmonov, P.Z. Generalized ternary Esterman problem for non-integer degrees with almost equal summands / P.Z.Rakhmonov // Mathematical Notes. 2016. - T. 100.   3. - C. 410 - 420.

[31] Vaughan R.C. Some remarks in Weyl sums / R.C.Vaughan // Coll. Math. Soc. Janos. Bolyani, Budapest 1981.

[32] Rakhmonov Z.H. Estimation of short trigonometric sums of G. Weil / Z.H. Rakhmonov, N.B. Ozodbekova // Reports of the Academy of Sciences of the Republic of Tajikistan. 2011. - T. 54.   4. - C. 257 - 264.

[33] Rakhmonov Z.H. Short trigonometric sums of G. Weil / .H. Rakhmonov // Scientific Notes of Orel University. Series of natural, technical and medical sciences. 2013.   6. part 2. - C. 194 - 203.

[34] Karatsuba, A.A. Theorem on replacement of trigonometric sum by a shorter one / A.A. Karatsuba, M.A. Korolev // Izvestiya RAN, Mathematical Series. - T. 71.   2. - C. 123 - 150.

[35] Arkhipov, G.I. Theory of multiple trigonometric sums / G.I.Arkhipov, A.A.Karatsuba, V.N.Chubarikov. - Moscow: Nauka, 1987, 368 p.

[36] Hua Lo-Gen Method of trigonometric sums and its applications in number theory / Hua Lo-Gen // - M.: Mir, 1964, 190 p.

[37] Estermann T. Proof that every large integer is the sum of two primes and square/ T.Estermann // Proc. London math.Soc., 11(1937). - P.

501 - 516.

[38] Karatsuba, A.A. Fundamentals of the analytic theory of numbers / A.A.Karatsuba // 2nd edition, M.: Nauka, 1983.

[39] Vinogradov, I.M. Method of trigonometric sums in number theory / I.M.Vinogradov // - M.: Nauka. 1980. 144 c.

[40] Vaughan R.C. On Waring's problem for cubes / R.C.Vaughan // J. Reine Angew. Reine Angew. Math., 1986, 365, - P. 122 - 170.

[41] Weil A. On some exponential sums / A.Weil // Proc. Nat. Acad. Sci., USA, 1948, 34, 5. -- P. 204 -- 207.

[42] Huxley M.N.. On the differences between consecutive primes / M.N.Huxley // Invent. math. 15. (1972). - P. 164 - 170.

[43] Rakhmonov, Z.H. Short linear trigonometric sums with simple numbers / Z.H. Rakhmonov // Reports of the Academy of Sciences of the Republic of Tajikistan. 2000. - T. 43.   3. - C. 27 - 40.

[44] Whittaker E.T. A Course in Modern Analysis, ch. 1. Basic operations of analysis / E.T. Whitteker, J.N. Watson // 2nd ed. Transl. from English, Fizmatgiz, M., 1963.

[Rakhimov, A.O. Asymptotic formula in the Esterman problem of the fourth degree with almost equal summands / A.O.Rakhimov // Reports of the Academy of Sciences ot the Republic of Tajikistan. 2015. - T. 58. 9. - C. 769 - 771.

[46] Rakhimov, A.O. Estimation of short trigonometric sums of G. Wweil of the fourth order in small arcs / A.O. Rakhimov // Reports of the Academy of Sciences of the Republic of Tajikistan. 2015. - T. 58. 8. - C. 674 - 677.

[47] Rakhimov, A.O. Short sums of G. Weyl and their applications / A.O. Rakhimov, N.N. Nazrubloev, Z.H. Rakhmonov // Chebyshev Collection. 2015. - T. 16. B. 1(53). - C. 232 - 247.

[48] Rakhimov, A.O. Short trigonometric sums of G. Weyl in the set of points of the first class / A.O. Rakhimov, N.N. Nazrubloev // Reports of the Academy of Sciences of the Republic of Tajikistan. 2014. - T. 57. 8. - C. 621 - 628.

[49] Rakhimov, A.O. On one additive problem with almost equal summands / A.O. Rakhimov, F.Z. Rakhmonov // Studies on algebra, number theory, functional analysis and related questions. ISSN: 1810-4134. 2016. 8. - C. 87 - 89.

[50] Rakhimov A.O. About one additive problem with almost equal summands / A.O. Rakhimov, F.Z. Rakhmonov // Proceedings of the international scientific conference "Modern problems of mathematics and its applications", dedicated to the 25th anniversary of the State Independence of the Republic of Tajikistan. Dushanbe, June 3-4, 2016. - C. 107 - 109.

[51] Rakhimov, A.O. Short trigonometric sums of G. Weyl in the set of points of the first class / A.O. Rakhimov, N.N. Nazrubloev // In: Algebra, number theory and discrete geometry: modern problems and

applications. Proceedings of the XIII International Conference devoted to the eighty-fifth anniversary of the birth of Prof. Sergey Sergeevich Ryshkov. Tula State Pedagogical University named after L.N. Tolstoy. L.N. Tolstoy. 2015.   - C. 245 - 246.

[52] Rakhimov A.O. About short trigonometric sums of G. Weyl / A.O. Rakhimov // Proceedings of the international scientific conference "Mathematical analysis, differential equations and number theory" dedicated to the 75th anniversary of Professor T.S. Sabirov, Dushanbe, October 29-30, 2015.   - C. 28 - 29.

[Rakhimov A.O. Estimation of short trinometric sums of G.Weyl of the fourth order in small arcs. / A.O. Rakhimov // Proceedings of the International Scientific Conference "Contemporary Problems of Mathematics and its Applications". Dushanbe, May 30-31, 2024.   - C. 148-150.

I want morebooks!

Buy your books fast and straightforward online - at one of world's fastest growing online book stores! Environmentally sound due to Print-on-Demand technologies.

Buy your books online at
www.morebooks.shop

Kaufen Sie Ihre Bücher schnell und unkompliziert online – auf einer der am schnellsten wachsenden Buchhandelsplattformen weltweit! Dank Print-On-Demand umwelt- und ressourcenschonend produziert.

Bücher schneller online kaufen
www.morebooks.shop

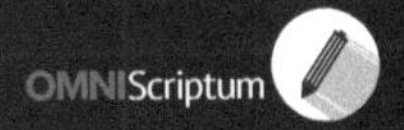

Printed by Books on Demand GmbH, Norderstedt / Germany